Close Encounters?

Science and Science Fiction

CLOSE ENCOUNTERS?

SCIENCE AND SCIENCE FICTION

Robert Lambourne, Michael Shallis and Michael Shortland

Adam Hilger
Bristol and New York

British Library Cataloguing in Publication Data

Lambourne, Robert
  Close encounters?
  1. Scientific knowledge related to science fiction   2.
  Science fiction related to scientific knowledge
  I. Title   II. Shallis, Michael   III. Shortland, Michael
  501

ISBN 0-85274-141-3

US Library of Congress Cataloging-in-Publication Data

Lambourne, Robert
  Close encounters?: science and science fiction/Robert Lambourne,
  Michael Shallis, Michael Shortland; with a preface by Paul Davies.
    198 p. 24 cm.
    Includes bibliographical references.
    ISBN 0-85274-141-3
    1. Science fiction—History and criticism.   2. Science in
  literature.   3. Literature and science.   I. Shallis, Michael.
  II. Shortland, Michael.   III. Title.
  PN3433.6.L36   1990
  809.3'8762—dc20   90-33934

Published under the Adam Hilger imprint by IOP Publishing Ltd
Techno House, Redcliffe Way, Bristol BS1 6NX, England
335 East 45th Street, New York, NY 10017-3483, USA

Typset by BC Typesetting, Bristol BS15 5YS
Printed in Great Britain by Butler & Tanner Ltd, Frome

CONTENTS

FOREWORD

I once asked a physicist colleague who it was that had first kindled his interest in science. 'Superman' was his instant reply. 'I owe a great debt to that guy.' Like many professional scientists, he could trace his passion for the subject back to some form of science fiction. In my case it was H G Wells and John Wyndham. Their stories fired my youthful imagination no less than all those carefully written accounts of relativity and subatomic particles. Science fiction is, in its relation to science fact, a unique genre, quite distinct in this respect from, say, historical romances or political thrillers. The reason for this concerns not so much the subject, but the style in which it is delivered, for science fiction is really a literary device for conducting a type of theoretical science, namely, the exploration of imagined worlds.

As a theoretical physicist my job is to construct mathematical models of idealized hypothetical universes, and then to investigate their properties. The science fiction writer has more scope. I am supposed to stay within the bounds of the accepted laws of physics, she or he doesn't. Nevertheless, the spirit of 'What if?' pervades both enterprises, and it takes a certain type of mind to appreciate that kind of approach to reality. There are, of course, many instances where science fiction novels have floated theories too speculative for the 'straight' science of the day, but which have subsequently turned out to be prophetic. Wells was a master at this. But I don't think the correctness of the science is as important as developing the sense of wonder, adventure and excitement—emotions experienced in the pursuit of real science.

Because many people's main exposure to science is through science fiction, the portrayal of the scientist, and the nature of scientific activity, is of crucial importance. All too often the characters are stereotyped and bear little resemblance to real scientists. I have yet to meet a real life Dr Frankenstein, let alone a Dr Who. Sadly, perhaps, scientists tend to be rather ordinary. What might be mistaken for eccentricity is usually social unease, common among people whose work demands dedication and long hours. And absent-mindedness is more often due to the distractions of an oppressive work load than genuine mental oddity.

And what of the science itself? Science fiction is incredibly variable in the accuracy of its science content. Compare the howlers in those lurid fifties B movies about alien invaders with the meticulous accuracy of *2001: A Space Odyssey*. Sometimes the science is used simply as a backdrop for a good old-fashioned yarn.

It is hard to see *Star Trek* as much more than a rerun of well-worn cowboy stories in a space-age setting. By contrast, Michael Crichton's *The Andromeda Strain* or Robin Cook's *Coma* take the science as their central theme, and deal with its consequences for humanity.

The best science fiction achieves more than mere entertainment. It tackles deep philosophical or ethical issues, and widens the reader's vision of the universe. It may also be designed to shock. I still carry the emotional scars of a childhood exposure to the horrors of *Quatermass* and, while I enjoyed both the book and the film versions of *Alien*, I agree with one film reviewer that the famous 'chestbusting' scene was a 'visual emetic'.

In spite of the evident popularity of science fiction, and its influence in shaping public attitudes to science and scientists, there have been very few serious studies of the genre and its social dimension. This book, from distinguished writers and academics, is a very welcome contribution to a neglected field. The authors address the place of science fiction in our society and the (often sloppy) treatment of science and scientists therein. There is also a robust discussion of pseudoscience, itself a growing (and in my view alarming) influence on the public perception of science.

Commercially, science fiction has an impressive track record, and is clearly here to stay, but as a literary genre it remains distinctly second class. Why? Perhaps it is because science itself is considered culturally inferior to the arts. Even as late as a century ago the arts and sciences formed an indivisible foundation for Western cultural life. But today, science, for all its acknowledged utility, is rarely appreciated as a cultural activity at all. If science is to be restored to its rightful place in our cultural heritage then science fiction writers carry a heavy responsibility.

Paul Davies
Newcastle-upon-Tyne
July 1989

PREFACE

Science fiction is booming. The space devoted to science fiction in bookshops and libraries has noticeably increased in recent years, science fiction films figure prominently in lists of highly profitable movies, television channels are presenting seasons of science fiction films with increasing regularity and advertisers are turning to science fiction for some of their most arresting and ingenious images. Ironically, this great boom in science fiction is happening at a time when science itself—particularly such pure science as astronomy and particle physics —is threatened by shrinking research grants and a decline in student interest, exacerbated by low morale and a shortage of science teachers. Environmental concerns and the ethical issues raised by genetic engineering, for example, have also made the public more sceptical of science; they no longer necessarily see it as an altogether beneficial enterprise. Not surprisingly, scientists, sensing danger in this altered climate, have become much more image conscious. Real efforts are now being made to improve the quality of science teaching at all levels and to increase the public awareness, sympathy and understanding of science. More and more scientists are becoming aware of the need to communicate the nature, value and significance of their work to the lay public. Greater and greater efforts are being made on many fronts to correct the widespread misconceptions about science and its practitioners. This book, we hope, will play a part in that process by examining the many ways in which images of science and the scientist are portrayed in science fiction. It will also play a part by illustrating just how useful and important science fiction can be in bridging the gap between the worlds of professional scientists and lay people. Science fiction has, we believe, a good educational role to play.

We hope this book will appeal to a wide readership. Certainly it should interest science fiction fans and students of film and popular culture, but more generally we also hope it will attract those with interests in the role of science in society and in the way that role is portrayed in the media. We are especially keen that it should be studied by scientists and teachers, particularly those who may not have read any science fiction since their youth, but who should nonetheless be aware of the complex picture of science that is presented by this popular genre and of its value in reaching a large (mainly young) audience. To achieve this wide appeal we have avoided the sort of afficionado's language and assumptions that would limit the readership to those who are already well versed in science fiction. Neither have we followed the purist's line in only dealing with written science fiction. We have treated the subject as it appears in the widest forms of popular culture, even though film and television presentations may not always have the same subtleties as the written word, nor the scientific detail. Our firm belief is that most people are exposed to science fiction through film and television and it is by coming to appreciate what those forms of the genre have to offer that the most representative view of the relationship between science and science fiction will be obtained. We hope that the book is self-contained and that it will provide even those who have never knowingly read a science fiction story with a broad overview of the genre and an understanding of its connections with science. At the same time, the extensive bibliography and filmography which follow the main text should provide new leads and fresh sources of insight for those who already have a long-standing interest in the subject.

The eight chapters into which this book is divided deal with aspects of the central theme from a number of different perspectives. The book does not offer a particular argument and therefore does not come to a specific conclusion; what it does, rather, is to present different approaches to the way science and science fiction interrelate. It is up to readers to reach their own views, in the light of their own experiences, as to where this relationship leads; we prefer to leave the matter open-ended. Each chapter was mainly the work of one author and, although we have attempted to ensure a reasonable uniformity of style, we have not tried to completely obliterate all signs of the diverse backgrounds and interests which we individually brought to the task of writing. Indeed, it is this diversity that, in our opinion, is one of the qualitites that distinguishes this book from others that have dealt with similar subjects. We have concentrated our attention on English language science fiction or works readily available in translation. The balance between books and films is fairly even-handed though there is a slight tendency to favour film since this is both numerically the more 'popular' medium and is the speciality of two of us. From the scientific point of view the emphasis is mainly on the physical sciences, but, as we argue in the text, this reflects a strong emphasis within science fiction itself. No attempt has been made to tackle the thorny issue of 'what is science fiction?' We leave the matter of what are often arbitrary demarcation lines to others and prefer to take the pragmatic view that if a work can be sensibly regarded as science fiction and if it is of relevance to this study then it should be included.

In order to provide some orientation, especially for the reader with little knowledge of the subject, the first chapter describes the origins of science fiction and picks out some of the main landmarks in its development. Many of these landmarks constitute a litany that is already well known to science fiction readers but we hope we have set them more firmly against their scientific background than is usually the case. This introductory discussion is followed by an examination of the way in which science enters science fiction. We have examined the role that it plays, the range and depth of the scientific ideas that are called upon, and the accuracy with which those ideas are presented, together with the question of where the science becomes imaginary. Chapter 3 continues this investigation by concentrating on a single theme, namely the role of time and time travel in science fiction. The tone of the discussion changes somewhat in the following three chapters, which concentrate mainly on a study of the science fiction films of the 1950s. These films have a particular importance, not just because they enjoy frequent reruns on television, but also because they are rich in images of scientists, their working environments and their personal and scientific goals. The three chapters deal in turn with the humanization and domestication of science by the movie industry of the 1950s; the representation of the scientist, whether as a genius, an evil monster or a regular guy; and, finally, with the cinema's treatment of the control of science and the integration of scientific ideology into social, political and traditional value patterns. Chapter 7 picks up this theme and examines the interaction between science and religion as presented in science fiction. As the chapter shows, that interaction is to some extent mediated by the outlook of scientists and the philosophy of science and therefore provides some insight into the wider issues of the role and location of science in society and the interchanges between science and theology. Chapter 8 provides a continuation of some of these topics as well as giving the book a suitably apocalyptic conclusion by examining ecological and environmental themes in science fiction, with an emphasis on both natural and man-made disasters and catastrophes.

As indicated earlier, the book ends with a bibliography for printed references, a filmography, giving production information (limited to director, screenwriter and producer) for each film cited in the text, and an index. We have adopted the convention of presenting book, journal and film titles in *italic script* and of enclosing the titles of articles or short stories in 'inverted commas'. The dates of works cited in the text are generally the dates of first publication or first release, though in the case of translations or works originating in classical times a specific modern edition is sometimes quoted. No attempt has been made to cite publishers for the great majority of science fiction works quoted. The rapid turn-over of these books combined with the differences between American and UK publishers made the provision of such information seem pointless.

Finally, we hope readers will find this book a source of insight into the rich material that science fiction contains. This genre may be classed as 'popular culture', and hence is all too readily dismissed because it is 'popular', but popularity does not necessarily mean of no value—it is still part of our culture. Indeed, science fiction is often the only arena within which serious and important ideas are expressed and examined. Cinema is frequently faced with the same criticism and yet film, as a medium, is just as rich a source as literature, music or the other arts. One trouble with popularity lies in the immediate accessibility of its offerings. The sifting of time reveals the classics. Bias against science fiction can also be located in public attitudes to science, which C P Snow discussed and dismissed as the 'two cultures' mentality.

We believe that this book will help many people re-evaluate science fiction and discover in it much that is thought provoking, serious and valuable, as well as finding it a wonderful source of entertainment and pleasure.

Robert Lambourne
Michael Shallis
Michael Shortland
1990

ACKNOWLEDGMENTS

Robert Lambourne acknowledges the invaluable assistance provided by the Science Fiction Foundation Library located at The North East London Polytechnic and extends special thanks to the Foundation's librarian Mrs Joyce Day.

Michael Shortland wrote his contribution to this book during August 1988, while he was a resident at the Bellagio Study and Conference Centre, Lake Como, Italy. He would like to thank the Rockefeller Foundation for enabling him to work in an ideal environment, the staff at the Villa Serbelloni for making his sojourn so agreeable, and fellow scholars in residence for their lively intellectual companionship.

Other thanks are due to Richard Fidczuk, Jane Gregory, Pamela Lambourne and the staff at the British Film Institute.

The stills reproduced in this book are from films originally distributed by the following companies, to whom thanks are due:

American Broadcasting Corporation, Edward L Alperson Productions, Allied Artists, Amicus Films, Cinerama Inc., Columbia Pictures, Bing Crosby Productions, Ealing Films, E.M.I., Los Altos Productions, Monogram Productions, M.G.M., George Pal Productions, Paramount Pictures, R.K.O., 20th Century Fox, U.F.A., United Artists, Universal International, Robert Wise Productions.

1

SCIENCE AND THE RISE OF SCIENCE FICTION

It all began with a wooden cow—at least that is the impression given by some historians of science fiction. Most modern investigators agree that science fiction first arose as a recognizable literary form some time during the nineteenth century, though the term 'science fiction' did not enter widespread use until the 1920s. However, few of those investigators can resist the temptation of looking back to earlier times for the precursors of science fiction. A favourite starting point for such discussions of the prehistory of science fiction is the ancient Greek myth of Daedalus and his son Icarus. That's where the wooden cow comes in.

Daedalus—a descendant of Hephaestus, the god of fire and metal-working—was renowned in ancient legend for his craftsmanship and inventiveness. But, despite his valuable skills, Daedalus was exiled from his Athenian homeland for the murder of his apprenticed nephew Talus, whose talents threatened to outshine his own. Taking refuge on the island of Crete, Daedalus found favour with King Minos and his Queen, Pasiphaë, by making articulated wooden dolls for the royal children and by creating a number of mechanical marvels. Unfortunately, Daedalus did not confine himself to such innocuous pursuits. During one of the King's absences he constructed an ingenious wooden cow in which Pasiphaë concealed herself while satisfying an unnatural passion she had developed for a sacred bull. A consequence of this exploit was the birth of a monstrous creature, half man and half bull: the Minotaur.

According to legend, when Minos discovered what had happened he made Daedalus build the Cretan Labyrinth, which he used as a hiding place for Pasiphaë's shameful offspring and as a prison for Daedalus. But the maze could not hold the artful craftsman for long. Daedalus freed himself from the Labyrinth with Pasiphaë's help, and then devised his most famous invention to make good his escape. Using wax and feathers he constructed wings so that he and Icarus could fly from Crete. The denouement of the story is well known. Despite being warned to follow his father and steer a middle course, Icarus was so exhilarated by the experience of flight that he soared too near the Sun. The wax holding the wings together melted, and Icarus plunged to his death in the sea:

> The unhappy father, a father no longer, cried out: 'Icarus!', 'Icarus', he called. 'Where are you? Where am I to look for you?' As he was still

calling 'Icarus' he saw the feathers on the water, and cursed his inventive skill.

[Ovid *Metamorphoses* Book VIII;
transl. May M Innes (1955)]

Although few would claim that the story of Daedalus qualifies as science fiction in the modern sense, the myth clearly contains many of the elements commonly found in science fiction. The central character is innovative, self-reliant, technically competent and, to some extent, at odds with society. The story involves a technological development that would have been well beyond the experience of its original audience, but not obviously impossible. Moreover, the action provides plenty of scope for igniting the 'sense of wonder' that is so highly regarded by modern science fiction readers. Even the more racy aspects of the plot would not be out of place in the setting of modern commercial science fiction.

The Daedalus myth is by no means the only product of classical antiquity in which portents of science fiction have been divined. Other works that have occasionally been claimed to be of relevance include Homer's *Odyssey*, Plato's *Republic* and Apuleius' *The Golden Ass*. (The last recounts the improbable adventures of a young man who is temporarily transformed into an ass.) In addition, two works by the second century Syrian satirist Lucian of Samosata are almost invariably included in historical accounts of science fiction because they both contain accounts of voyages to the Moon. In the first of these, *Icaromenippos*, the hero copies the example of Icarus, and uses wings to fly into the heavens, only to be returned to Earth by the gods. In the other, *The True History*, a ship is hoisted aloft by a waterspout and then propelled by winds to the Moon. Following the ship's arrival, the crew encounter a variety of bizarre creatures and become involved in a war between the King of the Moon and the King of the Sun over the colonization of Venus. After a short but eventful stay the seamen return to Earth where other fabulous adventures await them.

In spite of their undoubted literary merits, and their historical significance, none of these classical works can be of much interest to us in this study. They all lack the one ingredient of modern science fiction that will be our central concern: they show little sign of the influence of science as we currently understand it. For more visible indications of the effect of science on literature we must turn to a later epoch, when science itself had become a more highly developed and recognizable enterprise.

The phrase 'The Copernican Revolution' is used to describe one of the great upheavals in humankind's view of the Cosmos. As revolutions go it was a long drawn out affair. It started at some time in the early sixteenth century, when the Polish astronomer Nicolaus Copernicus called ancient authority into question by suggesting that the Sun, rather than the Earth, occupied the centre of the Universe, and it was not completed until a revised form of the Copernican system became firmly established in the seventeenth century. Most of the credit for the eventual victory of Copernicanism is usually given to the German scientist and cleric Johannes Kepler, Imperial Mathematician to the Holy Roman Emperor Rudolf II, and sometime assistant to the Danish astronomer Tycho Brahe. Following the latter's death in 1601, Kepler used the large body of superlative data amassed by the great observer to rationalize the Copernican system and to improve it by introducing elliptical planetary orbits in place of the circular ones that Copernicus himself had assumed. Kepler recorded his discoveries (including the three laws of planetary motion that are usually associated with his name) in a number of books and pamphlets such as *Astronomia Nova* (*The New Astronomy*) in 1609, and *Harmonicae Mundi* (*The Harmony of the World*) in 1619. Like many astronomers of his time, he was not simply a pure scientist; he was also expected to cast horoscopes and several of his works included detailed discussions of music, magic and mysticism. One book in particular, *Somnium* (*The Dream*), has come to be regarded as another forerunner of science fiction.

Somnium tells of a young Icelander, Duracotus, who returns home to his mother after spending some time in Denmark as a pupil of Tycho Brahe. The mother is delighted to learn of her son's astronomical knowledge, but already seems familiar with much of it thanks to her contact with a 'daemon' who has the ability to fly to the Moon during eclipses. The daemon is summoned and becomes the narrator of the story. It lectures Duracotus and his mother on the nature of the Moon and its relation to the Earth. Detailed descriptions are given of the mountains and seas of the Moon, the fortnight-long lunar nights, the appearance of the Earth, and what is seen during an eclipse. The numerous footnotes which Kepler added to the text during the last ten years of his life show that these descriptions were firmly based on the astronomical knowledge of the time. In addition, like many works of modern science fiction, *Somnium* speaks of those matters on which contemporary science was silent. It includes a description of snake-like lunar inhabitants and an account of the lunar weather.

The primary text of *Somnium* was probably written in 1593 and was revised in 1609, at about the time Galileo was making the first regular astronomical observations with a telescope. However, the book was not published until 1634, four years after its author's death, even though some copies had been in private circulation since 1611. According to one of the footnotes of the 1634 edition, the original text was written as 'an argument for the motion of the Earth, or rather a refutation of arguments constructed, on the basis of perception, against the motion of the Earth'. So, it was intended to demonstrate and propagate Copernican notions. This partly accounts for the delay in publication. The copious footnotes also make clear the deeply allegorical nature of the work, with Duracotus representing Science and his mother Ignorance. Comparisons between Duracotus' mother and Kepler's own mother, who was accused of witchcraft in 1615, also played a part in delaying publication and persuaded Kepler to add the explanatory footnotes. Lest it should be thought that books written with such an overtly didactic purpose cannot be regarded as science fiction, nor even as forerunners of science fiction, it should be noted that this kind of aim is not all that unusual in the field and will be encountered again.

Although Kepler was in the front rank of the 'New Astronomers', he was not alone in producing fiction that showed the influence of Copernicanism. Another account of lunar travel, *The Man in the Moone*, appeared in 1638. This one was by Francis Godwin, who became an Anglican bishop in 1601. It too was a posthumous work, published five years after its author's death, despite having probably been completed more than thirty years earlier. Unlike Kepler's story, this fanciful tale of a shipwrecked Spaniard who is towed to the Moon by migrating wildfowl was not at all concerned with scientific veracity and did not take a particularly pro-Copernican stance. Nonetheless, Godwin's story was widely read and influenced another cleric, John Wilkins, who later became the first Secretary of the Royal Society, in his influential non-fictional investigation *The Discovery of a World in the Moone, or, a Discourse Tending to Prove that 'tis probable there may be another habitable World in that Planet.* The first edition of Wilkin's book was published in 1638, the same

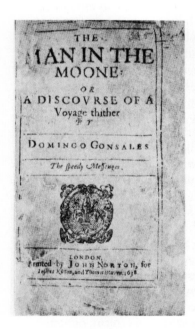 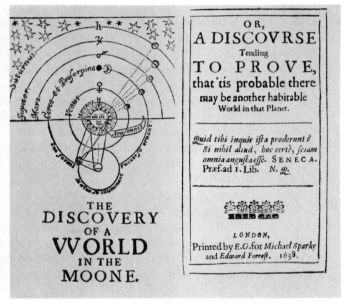

Science influenced by science fiction. Francis Godwin's *The Man in the Moone* (in which the hero is towed to the Moon by migrating geese!), and John Wilkins' mathematical and philosophical investigation *The Discovery of a World in the Moone*. (Reproduced by permission of the Trustees of the British Library.)

year that Godwin's story first appeared; it does not mention methods of reaching the Moon, but a discussion of that topic was added to the third edition of 1640 which explicitly mentions Godwin's unlikely proposal. On a more exalted plane, and with a somewhat shorter title, Milton's poetic epic *Paradise Lost* (1667) also contains an argument about the merits of the Copernican system. The discussion takes place in the Garden of Eden and is between Adam and the angel Raphael. It ends with Adam being warned to mind his own business:

> . . . heav'n is for thee too high
> To know what passes there; be lowly wise:
> Think only what concerns thee and thy being:
> Dream not of other worlds, . . .

[Milton *Paradise Lost* Book VIII]

Despite the angelic embargo, tales of voyages to the Moon and accounts of lunar inhabitants were highly popular throughout the seventeenth century. The growing scientific awareness of the Moon as a world in its own right, a world in some ways similar to the Earth, provided a basis for such tales and undoubtedly helped to boost their popularity. Several such tales were produced, though none of any great significance appeared in the fifty years that followed the publication of Cyrano de Bergerac's *A Voyage to the Moon* in 1657.

The scientific revolution of the sixteenth and seventeenth centuries reached its climax in 1687 with the publication of Isaac Newton's *Philosophiae Naturalis Principia Mathematica* (*Mathematical Principles of Natural Philosophy*). A number of literary works of the period show the influence of the development of science and that influence continues to be apparent in works of the eighteenth century. The exact nature of this influence has long been the subject of detailed academic investigations that have provided insights into works such as Daniel Defoe's *The Consolidator* (1705) and Jonathan Swift's *Gulliver's Travels* (1726). However, the extent to which such works can be fairly regarded as antecedents of science fiction is debatable, so we shall pass over them and move on to works that are more clearly related to modern science fiction.

In recent years, much has been written about the exact starting point of science fiction, and conflicting views have been expressed about the identity of the 'first' science fiction novel. Such a debate may well be sterile, and no more likely to come to a universally agreed conclusion than the search for a precise definition of science fiction itself. Nonetheless, the British writer and critic Brian Aldiss has for some time staunchly defended his view that science fiction really begins with the publication of *Frankenstein: or, The Modern Prometheus.*

The author of *Frankenstein* was Mary Shelley, the second wife of the poet Percy Bysshe Shelley. Mary was, as she described herself, 'the daughter of two persons of distinguished literary celebrity'. Her parents were Mary Wollstonecraft, the campaigner for women's rights, and William Godwin, the novelist and political theorist. Mary began writing *Frankenstein* shortly before her nineteenth birthday, two years after eloping with Shelley and some months before their marriage following the suicide of Shelley's first wife.

Fittingly perhaps, the book had two distinct incarnations: first, as a three volume work published in 1818, and then in its more familiar and somewhat reduced form dating from 1831. The circumstances that led to the writing of the story, including the role of contemporary scientific debates, are described in the preface to the 1831 edition, though there is some doubt about the accuracy of some of Mary Shelley's detailed recollections. In the summer of 1816 Mary and Shelley visited Geneva where they became neighbours of Lord Byron. They met Byron frequently, together with his physician Dr Polidori. The summer weather was poor, so the small party spent some time reading ghost stories and, following a proposal of Byron's, it was eventually agreed that each should write such a story. Polidori encumbered himself with an unpromising plot, and the two poets soon lost interest in the task, but Mary had her enthusiasm kindled by a vivid dream and by a conversation between Byron and Shelley. The conversation concerned, amongst other things, Galvanism, the principle to life, and an experiment then attributed to Erasmus Darwin in which a piece of vermicelli preserved in a glass case had reputedly begun to move with voluntary motion. The upshot of all this was Victor Frankenstein's monster, a creature assembled from parts of dead bodies and infused with 'the spark of being' by the scientist's 'instruments of life'.

The character of Frankenstein is probably best known today from films such as James Whale's 1931 adaptation which starred Boris Karloff. Most cinematic treatments provide an image of Frankenstein that differs markedly from Mary Shelley's original, but, as will be described in a later chapter, they have been undeniably effective in giving powerful expression to the demonic character of the mad scientist. The movies have also brought fame to Frankenstein's monster; it has become a modern media superstar, usurping the name of its fictional creator and escaping from books and films to make frequent appearances in adverts, pop videos and the like. Still, no matter how far modern representations of Frankenstein and his monster depart from the original, they nearly all preserve one aspect of the book—its Gothic style. The style originated in England with Horace Walpole's novel *The Castle of Otranto* (1764), and is generally characterized by horror, violence and supernatural effects set against a background of isolated castles and wild storms; the overall aim being to produce a sort of latter-day medieval romance. Like much early Gothic writing, *Frankenstein* provides a naturalistic explanation for its 'supernatural' elements and, more unusually, relates those explanations, albeit loosely, to the science of the day. In Aldiss' view, the Gothic nature of *Frankenstein* is highly significant. He regards science fiction as springing from the Gothic movement and asserts that science fiction is 'characteristically cast in the Gothic or post Gothic mode'. This view is not universally shared. Kingsley Amis, in his critical survey *New Maps of Hell* (1961), writes that 'The Gothic novel and its successors while all important to the ancestry of modern fantasy, scarcely prefigure science fiction'. However, Amis does make something of an exception of *Frankenstein*, which he describes as having 'a posthumous career of unparalleled vigour'.

Another name often brought into discussions about the origin of science fiction is that of Edgar Allen Poe. The case for Poe rests on a small number of short stories which use scientific or pseudo-scientific ideas, such as Galvanism and Mesmerism, as an integral part of the plot. These stories include 'Hans Pfaall—A Tale' (1835),

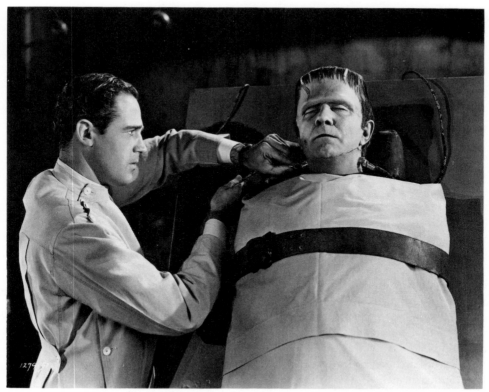

Frankenstein and his monster. A still from the 1943 reincarnation *Frankenstein Meets the Wolfman*.

which tells of an ascent to the Moon in a balloon. Poe's wide-ranging interest in astronomy, physics and mathematics is particularly clear in this story; it pays far more attention to scientific and technical details than anything written by Mary Shelley. The account of Pfaall's unparalleled adventures includes full descriptions of the balloon and the method used to maintain a breathable supply of air throughout the voyage, together with a mathematically based record of the changing appearance of the Earth and a good deal of sundry astronomical information taken from the American edition of John Herschel's *A Treatise on Astronomy* (1834). According to Poe, it was his study of this text that originally inspired him to write the story.

Within a few weeks of the initial publication of 'Hans Pfaall', newspaper reports started to appear in the *New York Sun* claiming that John Herschel, observing from the Cape of Good Hope with a large new telescope, had seen animals and vegetation on the Moon. These 'reports', which constituted what came to be known as 'The Great Moon Hoax', were actually the invention of Richard Adams Locke, one of the *Sun*'s reporters. Poe recognized them immediately for what they were, but many people were deceived. The most obvious effect of the hoax was to boost the sales of the *Sun* but Poe subsequently claimed that it had caused him to leave unfinished a second instalment of Pfaall's adventures which would have included descriptions of lunar life and conditions. If so, the Great Moon Hoax may have seriously undermined Poe's status as a science fiction author.

Whatever view is taken of Poe's significance in the history of science fiction, there can be little doubt of the importance of Jules Verne, who acknowledged Poe as one of his greatest influences. Verne, the son of a provincial lawyer, was born in Nantes in 1828. He was expected to follow his father into the law, but he yearned for a literary life. At first he tried his hand at plays and libretti and worked for a while in the Théâtre Lyrique in Paris. Eventually though, he discovered his bent for prose narrative and made that his future course.

An important event in Verne's career was his introduction, in 1850, to explorer and adventurer Jacques Arago, brother of the scientist François Arago. The house of Jacques Arago was a meeting place of scientists and travellers and it seems likely that it was there that Verne was stimulated to undertake the private study of science. He started with geography and physics and then moved on to mathematics. His reading was guided by a cousin who was a professor at the Ecole Polytechnique and the author of books on cosmography and mechanics. Verne worked hard, often starting at home at five in the morning and continuing in the reading room of the Bibliothèque Nationale, assiduously gathering facts and pursuing his literary ambitions.

He produced works for the theatre throughout the 1850s, though many of them went unperformed, and he interspersed them with the occasional short story. At the time of his marriage in 1857 he bought a fortieth share in a stockbroking company and began working in the Exchange, but he still continued to write. The turning point in his career came in 1863, when he submitted a semi-documentary account of exploration by balloon to the publisher P J Hetzel. At the time Hetzel was planning a monthly children's magazine to be called the *Magasin d'Education et de Rècrèation*, and Verne's work seemed just the mix of science and entertainment he needed. The book was accepted subject to modifications (which Verne completed within two weeks) and was published in 1863 as *Cinq Semaines en Balloon* (*Five Weeks in a Balloon*). In its revised form the book tells of the crossing of Africa from Zanzibar to Senegal in an innovative airship. Verne himself obviously regarded it as a work of great originality and high promise; he commented at the time:

> I have just finished a novel in a new form, a new form—do you understand? If it succeeds, it will be a goldmine.

> [Quoted from K Allott *Jules Verne* (1940)]

Verne had indeed found his goldmine. The book was an immediate success and resulted in a contract under which Hetzel was to be supplied with three volumes of the 'new form' each year, at a price of 1925 francs each. This was not a great deal of money at the time, but it was comparable to the sums paid to well established authors and it provided enough to live on. More favourable contracts, also with Hetzel, followed as Verne's fame grew.

For more than forty years, almost up to the time of his death in 1905, Jules Verne maintained a steady output of books as well as occasional articles, short stories and plays. Although he did not produce his first novel until the relatively late age of 35, by the end of his life he had written more than 60 books. The quality of the writing is sometimes criticized and occasionally mocked, but many of the stories are fondly remembered. Amongst his best known works are

Journey to the Centre of the Earth (1864)
From the Earth to the Moon (1865)
Around the Moon (1870)
20,000 Leagues Under the Sea (1870)
Around the World in Eighty Days (1873)
The Clipper of the Clouds (1886)
Master of the World (1904)

Verne's stories were published by Hetzel as '*voyages extraordinaires*'. Typically, they were packed with geographical and scientific facts. In several of them the technology of the day was extrapolated to allow the protagonists to experience fantastic adventures in exotic locations. This is clearly the case with the giant cannon used to launch men into space in *From the Earth to the Moon*. It is also true of Captain Nemo's submarine 'Nautilus' in *20,000 Leagues Under the Sea*, and of the 'Albatross', Robur's heavier-than-air flying machine held aloft by 74 vertical propellers in *The Clipper of the Clouds*. Although Verne's technical marvels were not always possible, he consistently showed attention to factual detail and demonstrated a desire to put his speculations on a scientific basis. When his own science was incapable of providing the necessary support for an idea, he was quite likely to call on the superior science of one of his characters. It is often pointed out that the space travellers in *From the Earth to the Moon* would be killed by the enormous acceleration at the moment they are launched from the giant cannon. This is a valid observation since the figures quoted by Verne imply an initial acceleration of more than 10 000 gees. Verne could have no real answer to such an objection, but he did realize that the initial acceleration was a problem and tried to take care of it with some novel (though unexplained) technology. Similarly, 'Nautilus' and 'Albatross' are powered by electricity, but, in Captain Nemo's words, 'it is not everybody's electricity.' This tendency to use science, even non-existent science, to lend plausibility to the fantastic, was a characteristic of Verne's work that was to influence many later authors. Verne may not have been the 'first' science fiction author, but he certainly helped to establish the style. His didacticism, his love of rationality, his taste for adventure and even the optimistic tone of his early novels were all to be copied by other writers.

The other figure that dominates the founding period of science fiction is that of the British polymath H G Wells. Born in Bromley, in 1866, Herbert George Wells came into the world at about the same time as Verne's fourth *voyage extraordinaire*.

Robur's flying machine 'Albatross' confronts a more primitive balloon in *The Clipper of the Clouds.*

His parents were servants turned shopkeepers who ran a small, and ultimately unsuccessful, china shop in Bromley High Street.

Shortly before Wells' fourteenth birthday, his mother left the chronically disorganized and underfunded family home to take up a job as a housekeeper. Wells' schooling was terminated and he was sent out to work as a draper's apprentice. He hated the job and caused so much trouble that his apprenticeship was not confirmed. This was just the first of a number of false starts that Wells had to endure. It was only after three hard years that his 'fifth start', as an assistant teacher and part time student at Midhurst Grammar School, put him on a profitable course.

The period surrounding Wells' birth and childhood had seen a number of remarkable developments in science. In 1859, Charles Darwin published his *Origin of Species*, and launched the public debate about evolution by natural selection that was to last for a generation. Between 1859 and 1864 James Clerk Maxwell used the power of mathematics to clarify and extend Michael Faraday's ideas about electricity and magnetism. This work culminated in one of the great unifications of physics, when Maxwell showed that light itself was an electromagnetic phenomenon—a self-sustaining pattern of electric and magnetic fields tumbling together through space. Even chemistry, a scene of enormous confusion, made great progress. The Russian, Dmitri Mendeleev, brought order to the elements in 1869 with his periodic table, though the long-running argument about the reality of atoms was to rumble on for many more years, until it was eventually settled by the triumphs of valency, kinetic theory and structural organic chemistry. Finally, as a background to everything else, the study of thermodynamics which had begun to emerge in the early 1800s had produced the concept of entropy and an awareness that, in a cosmic sense, the Universe was 'running down'. Of course, these developments were merely the highlights of a burgeoning scientific culture. There was a corresponding growth in technology and the need was perceived for an increase in the supply of scientifically trained teachers. As part of a programme to produce such teachers a system of training scholarships was introduced. Wells, a young man from an unprivileged background with a knack for learning and a burning desire to 'get on', was ideally placed to benefit from such a system.

After a year at Midhurst, Wells left to take up a scholarship at the Normal School of Science in South Kensington. At last, it semed, his future was assured. However, things were not to run smoothly. Wells' most important and successful year at the Normal School was his first, during which he came under the influence of Professor T H Huxley. Then and later, Wells regarded him as 'the greatest man I was ever likely to meet'. Huxley had become England's leading scientific spokesman and a powerful defender of evolutionary theory; he was also a brilliant and inspiring lecturer. Indeed, he provided such a clear and lasting impression of the power of evolutionary biology that Wells was to use the science long after as a yardstick for assessing other subjects. Wells described Huxley's course in biology and zoology as

> . . . a vivid, sustained attempt to see life clearly and to see it whole, to see into it, to see its inter-connexions, to find out, so far as terms were available, what it was, where it came from, what it was doing and where it was going.
>
> [H G Wells *Experiment in Autobiography* (1934)]

The physics course that occupied the first half of his second year left no such impression. It was taught by the ailing Professor Guthrie, who, Wells recollected, 'maundered amidst ill-marshalled facts', and by Guthrie's assistant C V Boys, 'one of the worst teachers who has ever turned his back upon a restive audience'. Almost equally bad was Wells' impression of the geology course that occupied the remaining eighteen months of his studies. Although Wells obtained a first class pass in zoology at the end of his first year, and second class passes in physics and elementary geology, he failed his third year examination in geology entirely and left the Normal School without a degree. Emerging at the age of 21 with poor qualifications and few prospects he found himself wondering 'what is to become of me *now*?'.

His destiny was, of course, to become one of the most successful authors of his day. His writings were to range from social comedy and utopian fantasy to panoramic surveys of history and polemical defences of his own political and educational beliefs. However, there can be little doubt that much of Wells' continuing fame rests on his 'scientific romances', of which the most famous are

> *The Time Machine* (1895)
> *The Island of Dr. Moreau* (1896)
> *The Invisible Man* (1897)
> *The War of the Worlds* (1898)
> *When the Sleeper Wakes* (1899)
> *The First Men in the Moon* (1901)
> *The Shape of Things to Come* (1933)

In addition to these and many other novels, Wells also wrote a large number of short stories. Ironically, though he was not primarily a technological extrapolator in the mould of Verne, it is probably for the predictions contained in some of these novels and short stories that Wells is best known. He is often credited with foreseeing such devices as the tank, the airscrew, the atomic bomb and even a form of genetic engineering.

The influence of Wells' scientific training can be seen in many of his works. His interest in evolution is certainly clear. *The Time Machine* tells of a trip into the far future, to a time when humankind has evolved into two distinct species—the innocent surface dwelling Eloi and the fearsome subterranean Morlocks. The theme of evolution is evident again in *the Island of Dr. Moreau*, though this time the 'evolution' is produced by the surgical intervention of Moreau in his attempt to bridge the gulf between Man and the beasts. Another indication of Wells' back-ground is the ability it gave him to keep abreast of scientific developments. In *The World Set Free* (1914), the book in which he described atomic warfare, Wells acknowledged his debt to Fredrick Soddy's scientific study *The Interpretation of Radium* (1909). It was from this work that Wells, a lifelong student of science, learnt of the emerging understanding of atoms and the possibility of releasing atomic energy. However, these examples simply provide superficial indications of a scientific training. Evidence of a deeper kind may also be found in his writings.

In her study *H. G. Wells, Discoverer of the Future* (1980), Roslynn Haynes draws attention to many deeper features of Wells' work that, in her view, reveal the

Publicity material for the film version of *The War of the Worlds*. Although the film did not accurately reflect the technology described in the book it did present a faithful apocalyptic vision.

influence of science. The topics she discusses include Wells' attitude to free will and determinism, his techniques of persuasion and presentation, and his mystical beliefs which, Haynes says, 'owe more to science than theology'. She also considers, at some length, Wells' approach to characterization. Not only is Wells' background to be seen in the numerous and often subtly drawn portraits of scientists but also in his unusually static and objective approach; 'as though viewing a specimen prepared for dissection'.

At first sight the careers of Verne and Wells seem strangely paradoxical. Verne, without any formal scientific training, wrote in a manner that emphasized factual accuracy and scientific credibility. Wells, who had received such a training, produced a kind of fiction that avoided scientific detail; he was more concerned with exploring ideas rather than setting them to work. The difference between the two men, as they themselves saw it, is clearly indicated by the following pair of quotations. The first presents Verne's view of Wells:

> I make use of physics. He invents. I go to the Moon in a cannon-ball discharged from a cannon. Here there is no invention. He goes to Mars

(sic) in an airship (sic), which he constructs of a metal which does away with the law of gravitation . . . Show me this metal. Let him produce it.

The second gives Wells' view of their relationship, or lack of one:

> . . . there is no literary resemblance whatever between the anticipatory inventions of the great Frenchman and these fantasies. His work dealt almost always with actual possibilities of invention and discovery and he made remarkable forecasts. The interest he evoked was a practical one; he wrote and believed that this or that thing could be done which was not at this time done. He helped his reader to imagine it being done and to realise what fun, excitement or mischief would ensue. Most of his inventions have, 'come true'. But these stories of mine . . . do not pretend to deal with possible things, they are exercises of the imagination in a quite different field. They belong to a class of writing that includes the Golden Ass of Apuleius, the True Histories of Lucian, Peter Schlemihl and the story of Frankenstein.

> [Both quoted from Franz Rottensteiner
> *The Science Fiction Book* (1975)]

Regardless of their own views, it is still perhaps Wells who was the more deeply driven by science. The crucial point perhaps is Wells' clearer understanding of the distinction between science and technology. It is Wells who shows the greater concern with overall order and general system. It is Wells who always strives to integrate his ideas into large-scale schemes based on some underlying structure. It is surely Wells who has been more imbued with the scientific spirit of his age, the spirit of evolution, of thermodynamics and of all those other grand scientific theories that were being developed alongside his writings. Despite their differences, or perhaps because of them, Verne and Wells stand together as the Galileo and Newton of science fiction. They may not have been the first of their kind but viewed from the present they are certainly the foremost. They were both prolific generators of ideas; inventing themes that other writers would develop and extend. Together, they helped to determine the form and nature of science fiction and they inspired innumerable people, particularly young people, with the desire to read more and to create more of the same kind of literature.

Although Verne and Wells fully deserve the credit they are given for moulding and popularizing science fiction, their emergence is just one indication of a more general development. The late nineteenth century saw not only an acceleration in the pace of technological progress with increasing industrialization and improving communications, but also a good deal of social progress. In Britain, the Forster Act of 1870 established local School Boards charged with making elementary education available to some of the one and a half million children aged between 6 and 12 who were, at that time, receiving no schooling. The supporting Acts of Sandon (1876) and Mundella (1880) had the effect of making such education compulsory and guaranteed an increase in the literacy and efficiency of the next generation of workers. In America, widespread elementary education had been a goal for some time. Under the leadership of Horace Mann and Henry Barnard, free elementary education under state auspices became standard practice, and by the mid 1890s more than thirty state legislatures had made school attendance compulsory. These changes in Britain and America were symptomatic of a new spirit that affected many other countries. The growth of literacy, affluence and leisure in a world where printing was becoming cheaper and distribution easier inevitably led to an increase in reading. Production of books, newspapers and magazines rose dramatically, creating new markets and new opportunities for aspiring authors.

One aspect of this changing scene was of particular importance in the development of science fiction: the birth of the 'pulp' fiction magazine in America in the mid 1890s. The pulps were so-named because of the kind of paper on which they were printed; a cheap, off-white rough-surfaced medium which yellowed quickly and became brittle with age. Those which specialized in fiction were aimed at a middle of the road market, more sophisticated than that of the earlier dime novels, but less affluent than that of the higher quality 'slicks' which were printed on well finished paper. The first of the all-fiction pulps was *Argosy*, published by entrepreneur Frank A Munsey. Following its launch in 1896 the magazine quickly attained a fairly static circulation of about 80 000, but that figure grew rapidly in the early years of the twentieth century reaching 500 000 around 1905. There was soon plenty of competition.

The early pulps, which included such titles as *All-Story, Pearson's Magazine* and *Blue Book*, were general fiction magazines. Costing from 10 to 25 cents, they typically consisted of 120 untrimmed pages in a 7 by 10 inch format with a lurid full colour cover. Science fiction and fantasy stories were fairly common, though they were rarely dominant. Most of the science fiction written for the pulp market was rather poor and is now largely forgotten or at least unread. However, one author, whose first novel was serialized in *All-Story* in 1912, remains popular today: Edgar Rice Burroughs, the creator of Tarzan. Burroughs began his writing career with *Under the Moons of Mars*, a tale of romance and high adventure in which heroic Earthman John Carter is mysteriously transported to Mars, where he does battle with six limbed, fifteen foot tall, green Martians to win the love and freedom of the beautiful princess Dejah Thoris. *Under the Moons of Mars*, which was later published in book form as *A Princess of Mars* (1917), was just the first of eleven John Carter adventures. In addition to the Martian series, Burroughs also wrote five stories about the exploits of one Carson Napier on Venus, a sequence of novels set in Pellucidar—a land situated more than 500 miles beneath the surface of the Earth—and numerous other works. More than twenty of these featured Tarzan, who made his first appearance in the October 1912 edition of *All-Story*.

From the scientific point of view the period coinciding with the growth of the pulps was extraordinarily exciting. X-rays were discovered in 1895 by Wilhelm Roentgen and radioactivity in 1896 by Henri Becquerel. Quantum theory was born in 1900 when Max Planck revealed his solution to the problem of black body radiation. Radium was isolated by Marie and Pierre Curie in 1902. Three years later, in 1905, Albert Einstein published his paper 'On the Electrodynamics of Moving Bodies' which laid out the special theory of relativity. Between 1906 and 1909 Ernest Rutherford and Hans Geiger performed the alpha-particle scattering experiments that prompted Rutherford to propose his nuclear model of the atom. According to Rutherford's model negatively charged electrons orbit a massive, positively charged central nucleus in much the same way that planets orbit the Sun. Within a few years the Danish physicist Niels Bohr had improved Rutherford's model by insisting that, unlike planets, the electrons in a nuclear atom were only allowed to orbit at certain well defined distances from the nucleus. Electrons could jump from orbit to orbit but they could not spiral from one orbit to another, nor could they establish new orbits in the forbidden region between one allowed orbit and the next; quantum theory did not permit it. Bohr's model of the atom was itself only an *ad hoc* stop gap destined to be modified and eventually superseded by more consistent models, but it was an important first step towards the modern quantum theory of the atom and it still provides the most advanced picture of atomic structure that most non-scientists are likely to meet. Finally, on top of everything else, came Einstein's theory of gravity: the general theory of relativity, which was presented in a paper to the Prussian Academy of Sciences in November 1915.

What effect did this ferment of ideas have on Edgar Rice Burroughs? Almost none. There is some evidence that he was familiar with the popular debate about the existence of canals on Mars that was in full swing at the time, but even his knowledge of this subject seems to have been no greater than that acquired from reading the newspapers. Burroughs was not the kind of author who could be expected to take an interest in serious scientific issues, and nor were many of his contemporary pulp writers. The stories they produced were escapist fantasies, churned out at high speed and low cost. Science—real science—was of little relevance to them and of equally little commercial value. However, as America became involved in the First World War and experienced a paper shortage, conditions were changing in the pulp market. In pulp literature, as in science, an age of specialization was dawning.

As the pulp market matured, magazines carrying a single type of story began to appear; westerns, romances and crime stories all had their own pulps. It was only a question of time before someone tried a science fiction pulp.

The person who finally took the risk of launching an all science story pulp was Hugo Gernsback. So great was the eventual impact of Gernsback's magazine that he is sometimes referred to as 'The Father of Science Fiction', and the best known annual awards for achievement in science fiction are called 'Hugos' in his honour. Gernsback was born in Luxemburg in 1884. He had a technical education which included a spell at the Technikum in Bingen, and went to make his fortune in the USA in 1904. His first business venture in America focused on the manufacture and supply of batteries. These efforts were short lived, but a sideline involving the import of scientific equipment grew into a radio business and caused Gernsback to produce a mail order catalogue. This led him into publishing.

Gernsback launched his first magazine in 1908. It was a popular science journal entitled *Modern Electrics*. The magazine changed its name to *Electrical Experimenter* in 1913 and to *Science and Invention* in 1920, by which time Gernsback was also publishing a companion magazine *Radio News*. Both magazines regularly ran science fiction stories. Science fiction had made its first appearance in a Gernsback publication in 1911, when extra copy had been needed to fill the April edition of *Modern Electrics* and Gernsback had responded by rattling off the first of twelve monthly instalments of *Ralph 124C41 +*, a romance of the year 2660. As the vocalization of its title suggests (one to foresee for one), *Ralph 124C41 +* was a story filled with forecasts of future technology. It was indicative of the kind of science fiction that Gernsback liked. In 1923 Gernsback published a special 'Science Fiction Number' of *Science and Invention* that included six science fiction stories (partly to get rid of a backlog that had built up), and in 1924 he sent out circulars inviting readers to subscribe to a new magazine he intended to call *Scientifiction*. Response must have been lukewarm at best because no such magazine appeared. However, in April 1926 he finally took the plunge and released the first issue of *Amazing Stories*, which he subtitled 'The Magazine of Scientifiction'.

Amazing Stories is still in print today, more than sixty years after its first appearance. Initially it featured a large number of reprints from the works of the founders of science fiction. The first issue included short stories by Wells and Poe and the first part of a serialized novel by Jules Verne. Stories by more typical inhabitants of the pulps were few. This pattern continued for some time as authors adapted themselves to the new market. The editorials that Gernsback contributed each month emphasized one particular feature of that market: its intention to take science seriously. This was underlined by the inclusion of a 'Discussions' section, in which readers could comment on issues raised by the stories, and by the introduction of quizzes. Questions such as: 'When is a planet in opposition?', 'What is the spectrogram of a star?', and 'Where do we find Widmanstatten figures in Nature?' enabled readers to check that they had managed to disentangle the scientific facts contained in the stories from their fictional wrapping. The slogan that appeared on the title page of each issue, 'Extravagant Fiction Today . . . Cold Fact Tomorrow', gave a clear indication of the kind of image Gernsback was trying to project.

Unfortunately, few of the stories that appeared in *Amazing* showed much sign of living up to the level of prophetic accuracy proclaimed by the slogan. A number of stories were written by people with scientific credentials and did avoid errors of fact, but even if the impossible was absent, the highly implausible was never hard to find. The standard of original writing was poor, the plots crude, the characters unrealistic, but still the magazine sold. A market for a specialized 'scientifiction' magazine existed.

In February 1929 Gernsback lost control of *Amazing* when he was forced into bankruptcy for late payment of debts. He claimed he was the victim of a conspiracy, paid his creditors 108 cents on the dollar and bounced back by launching two new magazines, *Science Wonder Stories* and *Air Wonder Stories*, within a few months. Faced with the necessity of competing with *Amazing* Gernsback decided to avoid the term scientifiction in his new publications. In its place he substituted 'science fiction' and is widely credited with putting that term into circulation, even though earlier examples of its use can be found. The new magazines, which were combined to form *Wonder Stories* in June 1930, continued to emphasize accurate science or at least the avoidance of inaccurate science. The educational aspect of science fiction was promoted and a 'Science Questions and Answers' department appeared. Lists of 'Associate Editors' who 'pass upon the scientific principles of all stories' were included in each issue and featured such well known luminaries as Lee deForest, the inventor of the triode, and the astrophysicist Donald H Menzel. Nonetheless, *Wonder Stories* never managed to equal the success of *Amazing*. Gernsback continued to publish *Wonder Stories* until the spring of 1936, but he was then forced to sell out. Apart from two other short lived ventures into the field, this ended Hugo Gernsback's active involvement with science fiction publishing.

Brian Aldiss, writing in *Trillion Year Spree* (1986), the Hugo-winning critical history of science fiction that he prepared with David Wingrove, points out that it is 'easy to argue that Hugo Gernsback was one of the worst disasters ever to hit the science fiction field'. Aldiss asserts that Gernsback's segregation of science fiction from other kinds of literature 'guaranteed the setting up of various narrow orthodoxies inimical to any thriving literature', and that the 'dangerous precedents he set were to be followed by many later editors in the field'. It certainly cannot be denied that most of the original stories carried by Gernsback's magazines were poor specimens that rarely lived up to the high standards of writing and scientific plausibility called for in the editorials. Nonetheless, for good or for ill, Gernsback did help to establish the style of the American pulp science fiction market and thereby influenced enormously the subsequent development of the field.

While Gernsback was changing science fiction, scientists were changing the Universe. 1917 saw the completion of the 100 inch reflector on Mount Wilson, a telescope that was to remain in a class of its own for the next thirty years and which was to become the veritable instrument of change. In 1918, Harlow Shapley, a young staff member at the Mount Wilson Observatory, announced the first results of his investigation into the size of our galaxy, the Milky Way. Shapley concluded that the centre of the Milky Way was 50 000 light years away, in the direction of the constellation of Sagittarius. By 1920, this pronouncement had led Shapley into his 'Great Debate' with H D Curtis. The debate concerned the size of the galaxy and the nature of the faint, cloud-like spiral nebulae that could be seen in various directions. Curtis maintained that the Milky Way was a much smaller galaxy than Shapley thought, and that the spiral nebulae were themselves galaxies beyond the boundaries of the Milky Way and not, as Shapley claimed, components of the Milky Way. As far as the spiral nebulae were concerned the debate was soon settled. In 1923, Edwin P Hubble, using the 100 inch reflector, showed that the brightest of the spirals, the Andromeda nebula, was approximately 800 000 light years away. This distance (which was actually a gross underestimate) was so much greater than the largest suggested diameter of the Milky Way that there could be no doubt that the spirals were external objects and that our galaxy was only one amongst many.

Nonetheless, Shapley had been substantially correct about the size of the Milky Way. Later work showed that the Sun was about 30 000 light years from the galactic centre; a somewhat smaller distance than Shapley's original estimate but much greater than the kind of figure Curtis had proposed. The 'Great Debate' had ended in a draw.

The development of galactic and extra-galactic astronomy in the years following the construction of the 100 inch telescope was reflected by a similar widening of horizons in science fiction. The August 1928 edition of *Amazing*, the issue that introduced Buck Rogers to an unsuspecting public, also carried the first instalment of a serial entitled *The Skylark of Space* by E E Smith. 'Doc' Smith's story was the first example of a particular kind of science fiction that would come to be called 'space opera', science fiction's equivalent of soap opera. Unlike earlier space stories, which had, almost without exception, confined themselves to the Solar System, space opera would routinely take the entire galaxy, if not several galaxies, as its setting. Titanic forces equipped with impenetrable shields and irresistible weapons would henceforth contend for the mastery of worlds. A small example of this heady brew is in order; the following extract is from Smith's 1948 revision of *Triplanetary* (first published in 1934):

> Torpedoes—non-ferrous, ultra-screened, beam dirigible torpedoes charged with the most effective forms of material destruction known to man. Cooper hurled his canisters of penetrating gas, Adlington his allotropic iron atomic bombs, Spencer his armour piercing projectiles, and Dutton his shatterable flasks of the quintessence of corrosion—a sticky, tacky liquid of such dire potency that only one rare Solarian element could contain it. Ten, twenty, a hundred were thrown as fast as the automatic machinery could launch them.

The Universe, in all its magnificent vastness, had finally squeezed itself between the covers of a pulp, but good writing and a plausible story-line had not. That development had yet to come.

Amongst those who contributed space opera to *Amazing* was John W Campbell Jr, a physics graduate from MIT and Duke University, who started his writing career while still a student. Campbell enjoyed considerable success as a writer, but he is mainly remembered as an editor. For more than thirty years, starting in 1937, he edited *Astounding Science Fiction* which dominated the American science fiction magazine market throughout most of the 30s and 40s. *Astounding* was launched in 1930 under the title *Astounding Stories of Super-Science*. It had a somewhat rocky start, soon changing owner, editor and title; by October 1933 it was in the hands of Street and Smith, a well established publishing company that brought it out on a regular monthly basis under the title *Astounding Stories*. Unlike Gernsback's magazines, *Astounding* had no educational pretensions and was soon the best selling science fiction magazine. Its more austere competitors, *Amazing* and *Wonder Stories*, were forced to become bimonthlies. In 1938, following Campbell's take-over, the title became *Astounding Science Fiction* and a new style soon began to be apparent.

Under Campbell's editorship *Astounding* demanded, and got, higher standards of writing, more logical plots and greater scientific accuracy. These improvements were brought about by a number of factors. One of the most important was the relative financial strength of *Astounding*, which paid its authors more promptly and at a higher rate than other magazines. This helped to make it a popular market with established authors such as Jack Williamson, Clifford Simak and E E Smith, and gave Campbell the freedom to be selective. Also important were Campbell's personal qualities—his ability to inspire his contributors, his gift for originating good ideas that others could develop into stories or serials, and his talent for spotting and nurturing promising newcomers. The combination of power and personality enabled Campbell to gather about him a group of writers, many of them young and technically trained, who had the ability and know-how to supply the kind of high-tech science fiction he required. Several of Campbell's protégés of the late 1930s are still best-selling authors today. This group includes Robert Heinlein, A E van Vogt and, probably best known of all, Isaac Asimov. Many authors of this generation have acknowledged the debt they owe to Campbell, none more so than Asimov:

> Campbell was a spider sitting in a web. To him came his fifty writers. He gave each one *his* ideas and watched for the sea changes that came back, and those sparked other ideas that he gave to other writers.
>
> He was the brain of the superorganism that produced the 'Golden Age' of science fiction in the 1940s and 1950s.
>
> He loved the role and, as far as I know, never abused it. He took no credit for himself. The three laws of robotics, the central idea of the *Foundation* stories, the plot of 'Nightfall' were all his, but although I always tried to make him take the credit for those things, he never would.
>
> [Isaac Asimov *Asimov On Science Fiction* (1984)]

The 'Golden Age' referred to by Asimov is widely regarded as having started in 1939. In that year the July edition of *Astounding* published van Vogt's first science fiction story, 'Black Destroyer', and an early work by Asimov entitled 'Trends'. The August edition introduced Heinlein with 'Life-line', and the September issue carried the first story of another newcomer, Theodore Sturgeon. A number of new magazines also appeared. There were eighteen by mid 1941, many of them making use of material that had been rejected by Campbell. These second rank magazines provided a foothold for yet more aspiring authors and helped to continue the boom into the early 1940s. They also made space for Ray Bradbury, the only major sf writer of the period whose career was not strongly influenced by Campbell. Bradbury's mytho-poetic evocations of childhood would soon bring him a fame that extended well beyond the narrow bounds of science fiction, but they would have been out of place in *Astounding*.

Golden Age or no Golden Age, the war years were a difficult time for science fiction. Although America was afflicted somewhat later than Europe, there was the inevitable paper shortage and also a lack of readers and writers. Many of the new magazines went to the wall, but *Astounding*, by the standards of the time, continued to flourish. Some of the stories reflected war-time preoccupations, including Cleve Cartmill's 1944 short story 'Deadline' which described atomic weapons in such detail that it prompted government agents to pay a call on *Astounding*'s offices. Campbell convinced them that the scientific content of the story was openly available in science journals and continued to print stories on atomic and nuclear themes. Indeed the dropping of the atomic bomb produced a vogue for holocaust and post-holocaust stories which became a major post-war theme.

The years immediately following the Second World War also saw the first appearance in America of works by the British writer Arthur C Clarke. Destined to be one of the superstars of science fiction, Clarke had spent the War as a technical officer in the RAF and had afterwards gone on to take a degree in maths and physics at King's College, London. By the time he graduated he was already the author of a number of stories and articles, including his celebrated 1945 contribution to *Wireless World* which pointed out the possibility of using artificial satellites in geostationary orbit as the basis of a worldwide communications network. His easy familiarity with science, his clear simple style, and his almost unequalled ability to activate the 'sense of wonder' in his readers, helped to establish Clarke as a gifted and popular writer. Later in his career, novels such as *Childhood's End* (1953) and *The City and the Stars* (1956) would make him one of

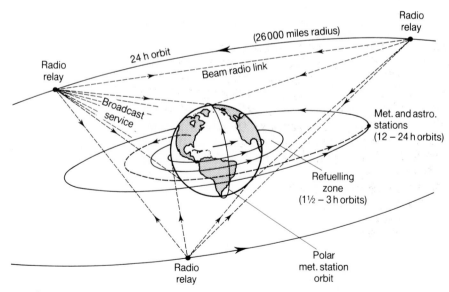

The use of geostationary satellites to provide a global communications network. (Reproduced from the revised 1958 edition of *The Exploration of Space* (1951), by Arthur C Clarke.)

the 'Big Three' alongside Heinlein and Asimov, and a wider fame would follow his involvement in the film *2001*, the screenplay of which he wrote in collaboration with Stanley Kubrick.

Opinions differ about the duration of the 'Golden Age' and about the true extent of its gilding; some maintain that it ended in the mid 1940s, others point out that it was nothing more than the coming of age of the rather juvenile magazine market that Gernsback had created. Whatever the truth of the matter, the fact remains that many works dating from this period, such as Asimov's original *Foundation* trilogy and Heinlein's *Future History* stories, are enduringly popular and are still readily available whereas most of the earlier products of the pulps are not. To the extent that science fiction is defined by what is found on the science fiction shelves of large bookshops, it must be accepted that the kind of stories which started to appear in *Astounding* around 1939 constitute a substantial part of that definition.

Amidst the factors that contributed to the creation of the Golden Age one has been omitted from the general discussion because it deserves special attention. It is the increasing scientific awareness of readers. This phenomenon was not entirely the result of improved education; in part the magazines themselves were responsible. In June 1936, before Campbell took over as editor of *Astounding*, the magazine started to publish a series of factual scientific articles under the collective title 'A Study of the Solar System'. Isaac Asimov recalls that the series 'was eaten up alive by the readers' and that he himself 'found it wonderful'. The series ran for eighteen consecutive issues and only ended when Campbell became editor. The author of this highly successful series was none other than John W Campbell. He continued to supply science articles after his take-over and eventually employed others to make them a regular feature.

Factual articles were not entirely new to the science fiction readers of the Golden Age. After all, *Amazing* had grown out of a popular science journal, and Gernsback's one page editorials in *Amazing* and in *Wonder* had frequently been devoted to scientific subjects. They had covered such topics as P W Bridgeman's Nobel-Prize-winning research into the properties of matter at very high pressure, Einstein's theories of relativity, and the nature of the degenerate matter in white dwarf stars. But Gernsback's editorials had been short and most of the other science content, apart from the quizzes and questions and answers columns, had been hidden in the stories. *Astounding*'s articles were significantly different; similar articles were to become a standard part of science fiction magazines, and their existence would lead to the creation of a substantial body of truly popular scientific literature. Isaac Asimov has been contributing such articles to one science fiction magazine on a regular monthly basis for more than thirty years. His essays are quite lengthy, and collections of them have formed the basis of some of his numerous non-fiction works.

The magazine which has published this series of articles by Asimov, *The Magazine of Fantasy and Science Fiction*, made its first appearance in 1949. It soon put an end to the period of *Astounding*'s unrivalled ascendancy. Familiarly known as *F&SF* it quickly acquired the reputation of being the most 'literary' of the magazines and of placing more emphasis on good writing than its rivals did. In 1950 it was joined on the newsstands by another highly regarded magazine, *Galaxy*, which came to be seen as the home of sociological and psychological stories, more concerned with the soft sciences than *Astounding*, and often written in a satirical or otherwise humorous style. *Galaxy*'s reputation for humour was given an enormous boost in 1952 by a serial entitled *Gravy Planet* that described a future in which commercialism had run amok. Written by Fred Pohl and C M Kornbluth, and subsequently published as a novel under the title *The Space Merchants* (1953), the story was hailed as a 'classic' and widely quoted as evidence of the maturity of science fiction. With the launch of *Galaxy* the pattern for the 1950s was set. Three magazines, *Astounding*, *F&SF* and *Galaxy*, would dominate. Their different images, different editors and different, though overlapping, teams of regular contributors, would widen the scope of science fiction and increase its humanity. However, what was to be a boom time for science fiction would not, in the long term, be so good for the magazines.

The era was one of diversification, development and growth. A large number of minor magazines were born, flourished for a while, and then died in the shadow of the majors. Mainstream publications started to take an interest in science fiction. A significant amount of science fiction began to appear in book form. At first these books were published by small specialist firms, usually established and manned by fans determined to preserve and disseminate the best of the stories from the magazines. Later, major publishing houses became involved. Anthologies appeared in greater and greater numbers and some authors began to write specifically for book publication. With the end of the decade came the first clear signs of the flood of paperbacks that would wash away the last vestiges of real power that the magazines retained. The magazines would survive, some of them at least, but their role would change and they would no longer be the sole source of new science fiction.

Few major new American authors emerged in the 1950s, though Frank Herbert and Philip K Dick published their first stories in 1952, and Robert Silverberg in 1954. Herbert would eventually write *Dune* (1965), a bestseller frequently voted 'best science fiction novel of all time' in polls of science fiction fans. Dick and Silverberg did not achieve the same kind of success with any individual work but both were enormously successful within the borders of the genre and, in Silverberg's case at least, well beyond.

Authors such as Asimov and Heinlein, who had become prominent in the 1940s, confirmed their positions with strings of new successes and with the publication in book form of the stories which had given them their initial popularity. Others, who had made less of a splash when they first appeared, grew to professional maturity. James Blish, sometime student of zoology at Rutgers University and an ex-US Army medical technician, was amongst them. It was in the 1950s that he wrote his highly successful 'pantropy' stories about the genetic modification of human beings to suit them to alien environments. He also produced three quarters of his *Cities in Flight* tetralogy in which the city of New York, equipped with an environmental recycling system and enclosed in a gravity screen cum force field, leaves the Earth to wander the galaxy in search of employment, profit and freedom. Another maturing talent was Poul Anderson, a physics graduate of Minnesota University, whose writing career began in the 1940s but only really took off in 1953. Anderson's prolific output of fiction ranges widely. Some of his works draw heavily on his technical background, including parts of his long running *Technic Civilization* series, others are almost pure fantasy.

The 1950s also saw a revivification of British science fiction. The genre had got off to a good start in Britain with H G Wells and those who followed immediately in his wake. It had also benefited from contributions by authors such as Rudyard Kipling, who wrote two short science fiction stories, and Sir Arthur Conan Doyle who fabricated a number of adventures involving Professor George Edward Challenger, starting with *The Lost World* in 1912 and ending with 'When the World Screamed' in 1928. British science fiction should have been given another powerful boost in the 1930s by the extraordinary works of Olaf Stapledon, but by this time something of a rift existed between British and American science fiction. Stapledon saw himself as writing in the same European tradition as Wells and was apparently unaware of the 'genre' represented by the American magazines until the late 1930s or possibly 1940. Stapledon's major works, *Last and First Men* (1930) and *Starmaker* (1937), are both epics of truly cosmic proportions. The former concerns itself with the evolution of mankind over the next 2000 000 000 years and describes the way changing conditions in the Solar System force our descendants to migrate to other planets; first Venus, then, a billion years hence, Neptune. *Starmaker*, for its part, recounts the journeying of a disembodied spirit as it moves through time and space, viewing alien species, watching the galaxies evolve towards telepathic consciousness, and searching for the creative cosmic spirit that called our universe and many others into being. The scope and richness of Stapledon's imaginings won him many fervent admirers, but somehow had relatively little impact on his fellow writers. John Wyndham, whose true name was John Wyndham Parkes Lucas Beynon Harris, started publishing science fiction in the 1930s under the name John Beynon Harris. He successfully aimed his early works at the American pulp market and became very popular, but initially showed little sign of his English origins and no sign of cosmic, as opposed to interplanetary, concerns. Similarly, Eric Frank Russell, who actually had the distinction of introducing Stapledon to the science fiction pulps, wrote and sold stories to American magazines that could easily have been passed off as the works of an American.

In the 1950s British science fiction authors again began to speak with a distinctive and significant voice. The case of Arthur C Clarke has already been mentioned. Clarke was the only prominent British writer of the period to acknowledge Stapledon's influences, but even he produced the bulk of his writing in a sort of international style that owed little to Britain. Indeed, Clarke left Britain in the mid 50s and settled in Sri Lanka which has been his home ever since. John Wyndham found a new voice with a more positively English accent. His decent, middle class heroes and heroines looked on while the world was ravaged by one

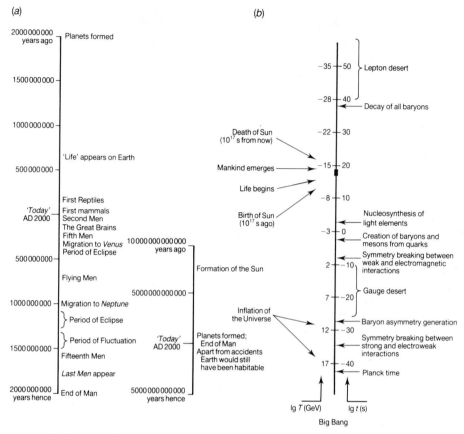

(a) Timescales showing the history of mankind and the cosmos from Olaf Stapledon's *Last and First Men*.
(b) The current scientific view of cosmic history from 'The Inflationary Universe' by Andre Linde, *Reports on Progress in Physics*, **47** 925 (1984).

catastrophe after another. First there was *The Day of the Triffids* (1951), then peril from beneath the seas when *The Kraken Wakes* (1953) and the post-nuclear problems of *The Chrysalids* (1955); little wonder that by 1959 Wyndham's space-faring Troon family were feeling *The Outward Urge*. Another Englishman who felt the outward urge in the late 50s was Cambridge astronomer Fred Hoyle. His fictional masterpiece, *The Black Cloud*, was published in 1957, shortly before his promotion from university lecturer in mathematics to Plumian Professor of Astronomy. Telling, as it does, of the efforts of a small group of mainly British astronomers to contact an intelligent cloud approaching the Solar System, the book carries the authentic atmosphere both of science and of Britain. The novel was obviously informed by its author's own position in the scientific establishment and its tone makes clear Hoyle's fascination with the methods as well as the results of science. No such influences could be said to bear on two other British writers who emerged in the 1950s: 'cult' author J G Ballard, and the 'grand panjandrum' of British science fiction, Brian Aldiss. Both have had a significant effect on the development of the genre, but neither has been primarily interested in its purely scientific content. More will be said of their role shortly.

Apart from books and Brits, two other factors had a major effect during the 1950s: films and television. From a modern perspective most science fiction films of the fifties seem technically crude. Many of them were created by writers and directors who knew little about science and less about the literature of science fiction. Some science fiction readers felt that the films contained little that had not already been prefigured in print and looked down on them as stereotyped throwbacks to an earlier and less subtle era. Others were simply glad to see images of tomorrow realized on the big screen. But whatever the feelings of the fans, films such as *Destination Moon* (1950), *The Day the Earth Stood Still* (1951), *Invasion of the Body Snatchers* (1955) and *Forbidden Planet* (1956) were widely popular and played an important part in creating and reflecting the public perception of science and the scientist at a time of rapid technological progress. Their role in this regard is central to the theme of this book and will be considered in detail in later chapters. For present purposes let it suffice to say that the films helped to broaden the appeal of science fiction in its most general sense by bringing an awareness of at least some of its themes to many non-readers.

The effect of television, prior to the first appearance of *Doctor Who* in 1963 and *Star Trek* in 1966, was less direct but may have been even more profound. One of the immediate consequences of the rise of television, it has been claimed, was to force a number of would-be authors to start their writing careers in science fiction, even though the field may not have been their own first choice. Isaac Asimov has argued that this situation was brought about by the narrowing of alternative markets for short stories that came in the wake of television. The validity of this argument is rather difficult to assess, but if correct it may go some way to explaining the anti-scientific tone that began to emerge in some works of the late 50s. Certainly, it is true that science fiction continued to be published in magazines, in substantial quantities, long after most other kinds of fiction had been more or less entirely swallowed up by the book trade. It is also the case that science fiction magazines have traditionally been a proving ground for aspiring authors and many of those who started their careers in the late 50s now reject or have simply transcended the label 'science fiction writer'. However, many influences have operated on those writers including, in some cases, their own very considerable success. Perhaps Asimov is right and television did compel writers who were not very interested in science, or even antagonistic towards it, to start writing science fiction, or maybe it just seemed like a good idea at the time. Whatever their reason, they certainly came, and having arrived introduced a significant new tone.

The relationship between science and science fiction in the 1950s was a rather distant one, except perhaps in the cinema, and even there only in a general cultural sense. It might have been expected that the determination of the structure of DNA in 1953, the development of transistor-based electronics throughout the decade, and the birth of the space age in 1957 would have had an immediate and dramatic effect on science fiction but such was not the case. Of course, the technically competent authors of the Campbell school were attuned to these developments and the facts were soon finding their way into the fictions, but those fictions were not really responding. In some ways the fiction seemed almost to be ahead of the facts. Genetic manipulation was already established as a popular theme, though the method employed was usually old fashioned eugenic breeding rather than genetic engineering. Asimov had started writing about robots with 'positronic' brains many years before Bardeen, Brattain and Shockley built the first transistor. Writers had largely explored the Solar System and were more concerned with the stars when Sputnik 1 went into orbit. What this outstripping of science by science fiction indicated was not any increase in the prophetic ability of science fiction writers, but rather a change of emphasis. The days of the Gernsbackian technological forecast were well and truly over; science fiction of the 1950s was still concerned with emerging technologies and with their likely effects on society but it was no longer preoccupied with the detailed hardware that would bring those technologies about. As Fred Pohl put it, 'a good science fiction story should be able to predict not the automobile but the traffic jam'.

By the end of the 1950s the network of science fiction writers and their readers, linked by books, magazines and conventions, had created in a universe fit for space cadets to live in. Science fiction could feed off itself, requiring only the occasional infusion of scientific information to stop it going off track. The technology for reaching the stars did not exist, but science fiction writers had invented it. They did not explain how it worked but they had a name for it—*hyperdrive*. Every reader knew that a jump through *hyperspace* would allow interstellar distances to be bridged in a matter of days. Every reader knew that a *blaster* was a weapon; its operating principles were unclear but its lethality was incontestable. Every reader knew that an *astrogator's* job was to guide a spaceship from star to star, plotting jumps through hyperspace and avoiding gravitational pitfalls. Every reader knew the universe. Some writers thought that universe was too small.

The 1960s and 1970s saw five major developments in science fiction that led directly to the state of affairs that still prevails today. None of these was primarily scientific so they will be dealt with fairly briefly.

First, there was the 'New Wave', a group of writers who rebelled against the standard universe of the American magazines and against their cramping editorial standards. Initially a British phenomenon, the New Wave was spearheaded by authors such as J G Ballard and Brian Aldiss, who published their work in the London-based magazine *New Worlds*. Although there had long been signs of discontent with the prevailing norms of science fiction, the New Wave only became really apparent after Michael Moorcock was installed as editor of *New Worlds* in 1964. Under his regime the magazine started to publish stories that took a darker, less optimistic and less technologically orientated view of the future. Many of the stories employed unconventional narrative techniques. Some made full use of the relaxed standards and increased artistic freedom of the sixties. Indeed, this happened to such an extent that those long-standing arbiters of national good taste, booksellers W H Smith and Sons, eventually decided that freedom had become licence and took the inevitable step of banning the magazine from their shops. Undaunted, *New Worlds* continued to publish, providing the closest thing science fiction ever had to an *avant garde*. In 1967 it even managed to secure a grant from the government-funded Arts Council.

In the context of science fiction, the term 'New Wave' was made popular in the US by author and critic Judith Merril in her anthology, *England Swings SF* (1968). Soon after its publication a number of American authors such as Harlan Ellison, Robert Silverberg, Samuel R Delany and Roger Zelazny were being described as 'New Wave', though, as William Sims Bainbridge points out in his sociological study *Dimensions of Science Fiction* (1986),

> . . . it would be a mistake to say that the new wave began in Britain and then washed across the Atlantic to the American shore. Individual authors in the United States were developing a new agenda for science fiction and would have become visible in any case.

The New Wave was never a great commercial success; many long-time fans of science fiction disliked its rejection of scientific optimism and saw its experimental and sometimes *outré* styles as pointless, or meaningless—or both. However, the freedom that it brought to science fiction, a freedom in both style and content, now tends to be viewed as beneficial since it helped to broaden the field and ensure its continued health.

The second major development of the 1960s and 1970s was the rise of fantasy. Around the middle of the period everybody seemed to be reading *The Lord of the Rings*, the trilogy of novels written by J R R Tolkien and originally published in the mid 50s. The scientific content of this monumental bestseller is nugatory but its influence on science fiction enormous. A cursory glance at the science fiction shelves of any large bookshop may not reveal a copy of Tolkien's masterpiece, but it will certainly indicate the number and variety of his imitators, the range of other fantasists, and the popularity of their works. In some shops what is usually referred to as 'fantasy' is carefully separated from science fiction but in most the two are allowed to intermingle. Inevitably the borderline between the two kinds of fantastic literature is often blurred and some authors write in either or both veins as the whim takes them.

A third influence was the changing role of the magazine alluded to earlier. Of the big three of the 1950s, one entered a terminal decline, one changed its name, and one survived pretty much unscathed. *Galaxy* lost circulation in the sixties and ceased publication in 1980. *Astounding*, still under Campbell's control until the time of his death in 1971, was rechristened *Analog, Science Fact/Science Fiction* in 1960 and continues to publish under that name. *The Magazine of Fantasy and Science Fiction* has also survived to the present day although its present circulation of about 50 000 is only about a half of *Analog*'s. Apart from *F&SF* and *Analog*, two other traditional science fiction magazines emerged into the 1980s: *Amazing*, still going after sixty-odd years and various vicissitudes though with a circulation of only 13 000 or so, and *Isaac Asimov's Science Fiction Magazine*, which was launched in 1977 and which currently sells about 80 000 copies a month.

The magazines have lost the power they held in the 1950s, but they still play an important role. Their editorials, book reviews, letters columns, timetables of forthcoming meetings and conventions, even their advertisements, all help to support a thriving network of fans and fan organizations which are amongst the most peculiar and striking aspects of science fiction. This network is further sustained by *Locus*, which bills itself as 'The Newspaper of the Science Fiction Field' and by many amateur fan magazines colloquially known as 'fanzines'. The whole subject of fandom is a very interesting one, but not of great relevance to this particular study. Most British readers of science fiction, even keen readers, may well never see any of the magazines mentioned above and may be totally oblivious of fandom. The well developed paperback market that now exists has almost totally replaced magazines as the source of science fiction for most readers, although the magazines remain a useful proving ground for new authors.

The growth of the paperback market that has just been mentioned was the fourth important development to affect science fiction. It started in the 1950s, but was mainly a phenomenon of the decades that followed. By the early 1970s, American

'Not Wanted On Voyage'. Cover for the fanzine *Crystal Ship* by Shep Kirkbride (1987). (Reproduced by permission of the artist and *Crystal Ship* editor John D Owen.)

publishers were issuing more than 200 science fiction titles per year, about 50% of them new, the rest reprints of established works. The majority of these titles were 'real' science fiction, as opposed to fantasy, but Tolkien was still selling at the rate of about a million copies a year so publishers were increasing the proportion of fantasy in their lists. This level of production represented something like 9% of the total fiction market.

The final factor, and the one mainly responsible for the enormous boom in science fiction that has taken place since 1960, was the success of sf in films and on television. In 1966, *Star Trek* made its first appearance. It was an expensive series to mount and failed to attract very large audiences. It was cancelled after three seasons, but its fans would not let it die. Reruns of the original 78 episodes have kept the series on television screens since production ceased in 1968, and have built a world-wide following for the adventures of Captain Kirk and the starship Enterprise. *Star Trek* has become a cult. Five *Star Trek* movies have been released and others will probably follow. The original TV screenplays have been recast as short stories and released in book form. The fans have met for conventions and have created a market for *Star Trek* paraphernalia and spin-offs of every description. A follow-up series, *Star Trek—The Next Generation*, has been the almost inevitable result.

On the film front, a major development was the release of Stanley Kubrick's film *2001: A Space Odyssey* in 1968. Widely regarded as the first film to have convincing simulations of outer space, the film set new and exacting standards for special effects in science fiction cinema. Films with equally stunning special effects and even more popular storylines followed, such as the 1977 blockbuster *Star Wars*, directed by George Lucas. Subsequent pictures in the *Star Wars* cycle and two films by Steven Spielberg, *Close Encounters of the Third Kind* (1977) and *E.T.* (1982), confirmed the public taste for big-budget escapist entertainment with high grade special effects and a science fiction motif. Many other films followed in the wake of *Star Wars*, notably Ridley Scott's terrifying *Alien* (1979) and his visually rich *Blade Runner* (1982), the latter based on the Philip K Dick novel *Do Androids Dream of Electric Sheep?* (1968). There was also Disney Studio's *The Black Hole* (1977) and de Laurentis' *Dune* (1984), both of which failed to do justice to the concepts that had inspired them.

Readers of science fiction do not expect films to offer the depth and variety that can be found in books and magazines. The nature of the film medium, particularly the emphatically *popular* film medium, makes it almost inevitable that action adventure stories broadly similar to the pulp fiction of the twenties and thirties will dominate. Nonetheless, there can be no doubt that the enormous success of films such as *Star Wars* and *Close Encounters* has done a great deal to popularize science fiction and make it a part of everybody's general cultural background, even if it has not done the same for science.

And so, we come to the present. Signs of science fiction are now to be seen everywhere. Childrens' toys, record albums, newspaper adverts, even rocket-shaped ice lollies and potato snacks in the form of flying saucers all proclaim the rise of

An example of 1970s special effects. The alien spacecraft from the film *Close Encounters of the Third Kind* (1977).

science fiction. The proliferation of this set of interconnected images relating to science, technology, other worlds and the future has been accompanied by an increase in the readership of science fiction. According to figures compiled by *Locus*, in 1989 American publishers released more than a thousand new 'science fiction' titles and nearly 650 reprints. Of the new titles about 65% were paperbacks; there were over 450 works of fantasy or horror and 279 'real' science fiction novels; the rest were mainly anthologies, single-author collections and novelizations. Science fiction was still booming even though the figures were slightly down on the previous year. The boom has brought about the age of the million dollar advance for the 'big-name' authors and has touched the career, not to mention the bank balance, of many others.

Writers such as Clarke and Asimov who had practically ceased to work on science fiction have been tempted back. Others, who began their careers in the 1960s and 1970s, have been carried forward by the floodtide. Amongst the most successful authors in this latter group are Ursula K LeGuin, one of the many female writers who have risen to prominence in recent years, and Larry Niven, who writes fast paced, technically literate fiction in a style that is a sort of cross between Arthur C Clarke and Robert Heinlein, with a dash of E E Smith thrown in. Niven is of particular relevance to this study since he is currently one of the most prolific and inventive exponents of scientifically based science fiction.

Niven's best known novel is probably *Ringworld* (1970); although written twenty years ago, it is not unrepresentative of his later work and provides a good example of the way in which science continues to influence science fiction. The novel deals with a ring-shaped artificial world, a million miles wide and two hundred million miles in diameter, built to supply its inhabitants with practically limitless amounts of living space. The story was inspired by physicist Freeman Dyson's suggestion that stars harbouring highly advanced civilizations might be so densely surrounded by artefacts that almost the entire energy output of the star would be absorbed and harnessed. Energy escaping from such a system would be coming from the relatively cool outer surface of the enormous 'Dyson sphere' enclosing the star and would therefore be mainly emitted in the infrared region of the electromagnetic spectrum. Dyson made this proposal in the context of the debate about the kind of astronomical sources that were most likely to be the abode of intelligent life. Until he put forward his idea, most searches for intelligent life had concentrated on Sun-like stars; the idea of examining infrared sources had not been seriously considered.

Younger authors have also emerged. Greg Bear, David Brin and Orson Scott Card have all achieved fame in the eighties. William Gibson, with his ultra high-tech 'cyberpunk' stories set in a world of unrestrained corporate competition, where major multinationals are quite likely to implant bombs in the brains of their important executives to prevent them from defecting to rival companies, has almost created a minor New Wave all of his own. The accompanying figure provides some insight into the current shape of science fiction by locating various authors with respect to the cardinal points of pure fantasy, pure hard science and pure New Wave. Of course, it would be historically unfair to classify as 'New Wave' authors who did not start their careers until the movement proper was dead and gone, but as Brian Aldiss has observed, 'In the States any writer with a freaky style became an honorary member of the New Wave', and it is in this extended sense that the term is used here.

Gregory Benford, who is both a noted science fiction author and a member of the physics faculty at the Irvine Campus of the University of California, has recently berated science fiction writers for losing touch with science. Writing in the November 1988 issue of *Amazing Stories*, he discusses John Updike's novel *Roger's Version* (1987) which touches on a number of scientific issues including the inflationary scenario for the origin of the Universe that was developed in the early 1980s by Alan Guth and others to augment and extend the Big Bang theory. Benford declares that 'It's an embarrassment to the SF field—*we* should have produced work at this level, about such deep and science-fictional issues', and he goes on to ask 'Why hasn't recent SF echoed the sense of process that is emerging in even the hardest of the physical sciences? . . . Why didn't an SF author take up the issues in *Roger's Version*?' Benford concludes by suggesting that '. . . we writers may be doing less homework. . . . Maybe we should start hitting the books again'.

Are Benford's criticisms valid? Have science fiction authors been neglecting to keep abreast of scientific development or at least failing to reflect those developments in their stories? Perhaps so, or maybe the situation is somewhat more complicated. As the next chapter will show, there are still some authors, including Benford himself, who are well informed about modern science and who delight in working recent discoveries and developments into their stories. At the same time, a

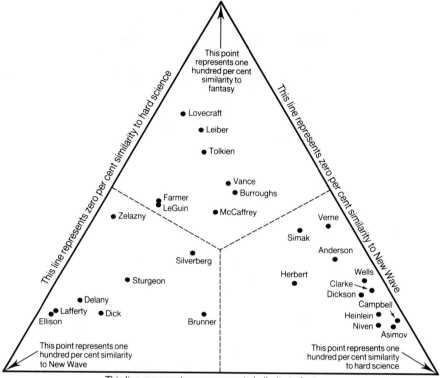

The 'shape' of science fiction as perceived by the fans, reprinted from an article of that title by William Sims Bainbridge and Murray Dalziel in *Science-Fiction Studies* volume 5, part 2 (1978). (Reproduced by permission of the editor.)

number of others, possibly the majority, find that science itself is of little relevance to their work. Perhaps this is simply a sign of the maturity of commercial science fiction. Thanks to the boom, the market-place is now well developed and highly diversified, and there is plenty of room for both the scientifically oriented author and the non-scientist. If the effect of this has been to widen the gulf that has always existed between science and science fiction some may feel that it is a positive development since it will reduce the risk that an inadequate image of science is taken seriously. Others may feel that such a development is wholly bad since it erodes a major link between technical and popular culture and increases the distance between the scientific community and the wider civilization of which it is a part.

2

THE SCIENCE IN SCIENCE FICTION

How does science enter science fiction? The very name *science* fiction seems to imply that it must be there, at least at the hard-science end of the field, the end occupied by authors such as Clarke, Asimov, Heinlein (in his younger days) and Niven. Of course, the science is not of the kind found in textbooks, at least not any longer. In the late 20s and early 30s Hugo Gernsback strongly promoted the idea that science fiction should be a vehicle for science education, an idea happily accepted by some of his readers though never fully reflected in the stories published in his magazines. 'Gernsback's fallacy' as it has been called was soon dispelled by the commercial success of rival publications that had little or no interest in educating their readers. Post-Gernsback science fiction does not deliver scientific information in great indigestible slabs, but it does make use of a wide range of scientific ideas. As new scientific concepts are developed there is occasionally something akin to a race between certain authors to incorporate them into stories.

Nowadays scientific ideas enter science fiction in a variety of ways, some of them quite subtle. Indeed, part of the art of writing hard science fiction is to introduce technical concepts with which the reader may be unfamiliar without appearing to lecture and without being boring. Some attempts at doing this are more successful than others. Here is an uncharacteristically bad piece of writing by the normally sane and reliable James Blish, taken from his 1973 novel *The Quincunx of Time*:

> "In a nutshell," Weinbaum said, "ultrawave is radiation and all radiation in free space is limited to the speed of light. The way we hype up ultrawave is to use an old application of wave-guide theory, whereby the real transmission of energy is at light speed, but a quasi-imaginary thing called phase velocity is going faster. But the gain in speed of transmission isn't much. By ultrawave, for instance, we get a beamed message to Alpha Centauri in one year instead of nearly four. Practically, that's not a very useful gain, even over that short distance. We need *speed*."
>
> "Can't it be made to go any faster?" she said, frowning.
>
> "No. Think of the laser pipe between here and Centaurus III as a caterpillar. The caterpillar himself is moving along quite slowly, just at the speed of light. But the pulses, the waves of contraction which pass along his body, are going forward in the same direction faster than he's going as a whole—and if you've ever watched a caterpillar you'll know that this is in fact the case. Now if the caterpillar is endless—tail on Earth, head on

Centaurus III—and we impose pulse modulation on those waves, we can get the message carried by the modulation there faster than the caterpillar himself would have gotten there."

The passage continues with Weinbaum explaining that those increases over light-speed transmission that have been achieved have only been made possible by the fact that the pulses travelling along that poor 'hard-working caterpillar' were carrying information and not energy, 'that remained constant'. The disquisition ends with Weinbaum asking his (predictably) female interlocutor "Clear so far?" to which she replies, with a sentiment familiar to many students of higher physics, "I think so. When I don't understand, I'll whimper".

The use of the caterpillar to try to get over the subtle but crucial difference between *group velocity*—the speed of energy propagation represented by the movement of the caterpillar as a whole—and *phase velocity*—the speed of the waves of contraction that pass along the caterpillar's body—is rather clever, but the overall impression is one of confusion. Perhaps this is deliberate. Blish was probably aware that the existence of phase velocity does not provide a means of sending signals at super-luminal velocities and was just using an explanation that was no explanation to blind his readers with science. H G Wells would have probably called this kind of thing *scientific patter*; he used it himself in his seminal novel *The Time Machine*:

"How did you come to travel through time?"
"By dimensional quadrature," Macklin replied.

Scientific patter can still be found in science fiction books, though it is rarer than it once was. It uses some of the flesh of science—the words and the jargon—but it completely ignores the blood and bones of the subject. It reveals nothing of the way scientists work and think and only a little of the way they talk. Scientific patter is not really regarded as 'respectable' by the hardest of hard sf authors and in any case it has been rendered obsolete by another tool of the science fiction writer—*imaginary science*.

Imaginary science is a term used to describe the network of scientific ideas, principles, laws, and hypotheses that authors imply as a background to their stories. Broadly speaking imaginary science is any kind of imagined 'science' that might plausibly be conceived as arising out of the continued progress of known science. An excellent example of imaginary science is the field of 'psychohistory' which provides the motivating spirit behind Asimov's *Foundation* series. Here is the definition of psychohistory as quoted from the Encyclopaedia Galactica (116th Edition) in the novel *Foundation* (1953):

PSYCHOHISTORY—... *Gaal Dornick, using non-mathematical concepts, has defined psychohistory to be that branch of mathematics which deals with the reaction of human conglomerates to fixed social and economic stimuli. . . . Implicit in all these definitions is the assumption that the*

human conglomerate is sufficiently large for valid statistical treatment.
The necessary size of such a conglomerate may be determined by Seldon's
First Theorem which . . .

A further necessary assumption is that the human conglomerate be itself
unaware of psychohistoric analysis in order that its reactions be truly
random. . . . The basis of all valid psychohistory lies in the development of
the Seldon Functions which exhibit properties congruent to those of such
social and economic forces as . . .

ENCYCLOPAEDIA GALACTICA

This is imaginary science writ large: an entire field of scientific endeavour with a mathematical underpinning, well defined limits of applicability and a chief practitioner, Hari Seldon. With a definition like this, quoted from such a respectable source, psychohistory sounds as real and as intimidating as quantum electrodynamics or non-equilibrium thermodynamics.

To some extent imaginary science has taken on a life of its own and has become the common intellectual property of a number of science fiction writers and their readers. The way in which terms such as hyperspace, blaster and astrogator had acquired commonly accepted meanings by the 1960s was mentioned in the last chapter, and many other examples of standard science fiction jargon could have been quoted (anti-gravity, force screen, teleporter, etc). The unspoken presumption that such devices are the natural outcome of normal scientific progress is so much a part of the fabric of science fiction that it does not need to be mentioned, but it clearly implies a substantial body of imaginary science. The unquestioned and unexplained use of imaginary science in this sense assists the writers of fast moving adventure stories by saving them the time and trouble of having to introduce concepts that are probably already familiar from the writings of others. Occasionally authors may impose a certain superficial individuality on imaginary science by calling hyperspace by some other name—'subspace' for instance, or 'ultraspace' or 'infraspace'—but the principle remains the same. Of course, some writers choose to explore such concepts in greater detail; indeed, they may even go as far as relating hyperspace or one of its analogues to the space–time 'wormholes' that are a respectable part of relativity theory. Effects of this kind, when taken far enough, cease to be examples of imaginary science altogether, but few stories go to these lengths.

Imaginary science is probably the most ubiquitous device used by science fiction authors to achieve the 'willing suspension of disbelief' on the part of the reader that is essential if their sometimes ridiculous sounding plots are to be at all convincing. It is the absence of the rich background implied by imaginary science that so often makes plot summaries and verbal accounts of favourite stories seem silly and hollow. To continue with the 'body of science' analogy used earlier, imaginary science uses the skeleton of science but, apart from the odd finger or toe, ignores the flesh and blood. It is an elaborate but hollow sham, using the meta-structure of science—the history, the philosophy, the framework of interlocking ideas, and the associated technology—to convince the reader of the truth of a fictional reality no matter how improbable it may sound. It is interesting that a hoax of this kind, willingly entered into by reader and writer alike, can be so effective. It tells us much about modern science that its form is so easily portrayed even when its content is absent.

In place of imaginary science some authors occasionally use *pseudoscience*. The term pseudoscience is usually taken to cover activities such as ufology (the study of flying saucers), astrology and a whole variety of 'psychic' sciences. Those who indulge in these activities may take them very seriously, they may even be scientists themselves and approach their subject in a rigorously scientific manner, but nonetheless, a sceptical scientific community will not accept them, even though it may not dismiss them out of hand. What pseudoscience offers an author is essentially identical to what imaginary science has to offer—the chance to invest a story with an instant background, an extra dimension, almost as complex and extensive as the background provided by science itself.

Some good examples of the use of pseudoscience are to be found amongst the small group of novels in which magic is used as an extension to science. A case in point is Larry Niven's *The Magic Goes Away* (1978), in which it is suggested that magic was at one time a natural physical process, but one that drew its energy from a sort of fuel called *manna*. When the extensive use of magic finally exhausted the world's supplies of manna the possibility of performing magical acts also ceased to exist. An even better example, perhaps, is *Magic Inc.* (1940) by Robert Heinlein. This is a novel replete with scientific references from an author with an engineering background who was in the vanguard of Campbell's *Astounding* revolution. The applied magic that Heinlein envisages is put to such useful purposes as the creation of 'magic' food that can be eaten and enjoyed, but which disappears after an hour or two leaving the eater no heavier, and ready for more.

By its very nature, pseudoscience is not particularly relevant to this study but it does serve to emphasize the lesson of imaginary science, that in trying to get the reader to suspend his disbelief, the form of a subject is often as important as its content. In the context of fiction the use of magic with its hierarchy of witches, warlocks and wizards, its spells and potions and its master practitioners such as Merlin and Gandalf, is just as valid as the use of quantum theory with its postgraduate students, research fellows and professors, its wave equations and uncertainty relations and its founding fathers Schrödinger, Heisenberg and Dirac.

Which brings us to science itself—the real thing. How does real science enter science fiction? Before answering this question let us look at the related issue of the *context* in which science arises in science fiction. What sorts of stories make use of factual science? We have already seen that science is of almost no relevance to fantasy, and of relatively little use to 'New Wave' writing which tends to be more concerned with the impact of science than with the science itself. It is therefore hard sf, the third of the traditions recognized by Bainbridge in his *Dimensions of Science Fiction*, that will mainly concern us. This is hardly surprising since a definition of hard science fiction (quoted by Bainbridge and attributed to L David Allen) is 'science fiction in which the major impetus for the exploration which takes place is one of the so called hard, or physical, sciences, including chemistry, physics, biology, astronomy, geology and possibly mathematics, as well as the technology associated with, or growing out of, one of those sciences'. Bearing this in mind, let us try to categorize sf stories in such a way that the role of science becomes clear.

Several attempts have been made to analyse sf on a thematic basis. The most thorough and carefully thought out of these is probably that of *The Encyclopedia of Science Fiction* (1979), prepared under the General Editorship of Peter Nicholls, which includes 175 theme entries. This listing is, if anything, too complete for our purposes: it repays careful scrutiny, but it hardly provides an easily assimilable

Table 1 Major themes in science fiction.

Themes and sub-themes	Examples
Space exploration and the military industrial complex	
Space travel and exploration	*Star Trek*
Colonization and exploitation of other worlds	*The Martian Chronicles* (Ray Bradbury)
Warfare and weaponry	*Star Wars*
Galactic empires and space opera	The *Lensman* series (E E Smith)
Aliens and monsters	
First contacts and other encounters	*Close Encounters of the Third Kind*
Biologies, environments and societies	*A Martian Odyssey* (Stanley G Weinbaum)
Invasions and aliens among us	*War of the Worlds* (H G Wells)
Artefacts and technologies	*Ringworld* (Larry Niven)
Future or alternate human societies	
Alternative histories	*The Man in the High Castle* (Philip K Dick)
Utopias and nightmares	*1984* (George Orwell)
Cities and cultures	*The City and the Stars* (Arthur C Clarke)
Transport, communications and technology	*The Stars My Destination* (Alfred Bester)
Cataclysms and dooms	*The Day of the Triffids* (John Wyndham)
Sex and cultural taboos	*Stranger in a Strange Land* (Robert A Heinlein)
Religion and philosophy	*Last and First Men* (Olaf Stapledon)
Men and Supermen	
Mutants, prodigies and symbiotes	*The Invisible Man* (H G Wells)
Telepathy, psionics and ESP	*Slan* (A E van Vogt)
Medicine and bionics	*$6 000 000 Man*
Robots, computers and gadgets	
Robots, androids and gadgets	*I, Robot* (Isaac Asimov)
Computers and cybernetics	*Colossus* (D F Jones)
Extra dimensions	
Time travel and lost worlds	*The Time Machine* (H G Wells)
Parallel worlds and extra dimensions	*Worlds of the Imperium* (Keith Laumer)

overview of the major themes. More simply grasped is the list of themes printed in table 1, which has been adapted from *The Visual Encyclopaedia of Science Fiction* (1978) edited by Brian Ash. A good many science fiction books, especially those of the 'harder' variety, may be fitted into one or other of these categories, as indicated. Books with a more complicated thematic structure can be seen as the result of interweaving several of the themes.

The abbreviated thematic listing of table 1 provides some valuable insights: certain branches of science are clearly relevant to some of the topics listed in the table. But still, the role that science plays in sf is not entirely clear. Perhaps more light can be shed on this aspect of science fiction by considering stories in the context of the following six-fold classification, even though it may over-simplify the situation to some extent.

(1)   *Using science to provide a description of a real but relatively unfamiliar environment, the description being based on scientific information available at the time of writing.*

The most characteristic stories of this kind are those set on planets or other bodies in the Solar System. Typical examples are *Heart of the Comet* (1987) by Gregory Benford and David Brin, which tells of a mission to explore and colonize Halley's comet, and many of the short stories written by Arthur C Clarke, including his award-winning novella *A Meeting with Medusa* (1971) in which a robot explorer floating in the atmosphere of Jupiter discovers native life. Clarke's highly successful 1961 novel, *A Fall of Moondust*, also falls into this category, though its story of a 'moonboat' that sinks into a sea of lunar dust has been rendered obsolete by the scientific finding that such dust seas do not exist. A similar comment applies to Alan E Nourse's short story 'Brightside Crossing'. This gripping tale of an attempt to cross the illuminated hemisphere of Mercury in specially adapted vehicles was written in 1951, at a time when it was still widely thought that Mercury rotated on its axis and revolved around the Sun in the same period of time, 88 days, thus keeping one side in permanently fierce sunlight and the other in perpetual shade. This was shown not to be the case in 1965 when radar observations of the planet indicated a 60 day rotation period. Nonetheless, 'Brightside Crossing' remains a little masterpiece of scientifically based science fiction. Many other stories of this general kind can be found in the collection *The Science Fictional Solar System* (1980) edited by Isaac Asimov, Martin Harry Greenberg and Charles G Waugh.

(2)   *Using science to provide a description of an imaginary environment that is as consistent as possible with established facts and principles.*

Included in this category are many of the stories set in habitats in the immediate vicinity of stars other than the Sun. Most of these 'habitats' are imaginary planets carefully designed by the authors who created them to stand up to a reasonable level of scientific scrutiny. The most famous example of this art of 'worldsmithing' must surely be the planet Mesklin which features in Hal Clement's novel *Mission of*

(a)

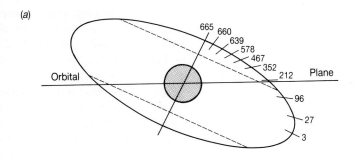

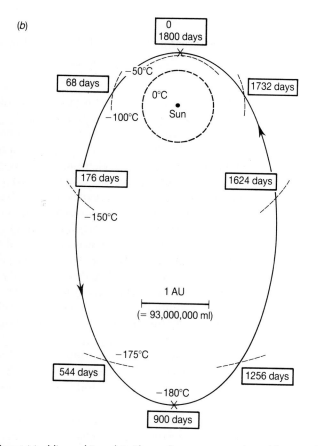

The planet Mesklin and its orbit. These diagrams reproduced from Hal Clement's article 'Whirlygig World' give some idea of the detailed thought that went into the worldsmithing of *Mission of Gravity*. (Reproduced by permission of Hal Clement.)

Gravity (1953). Clement, a high school chemistry teacher and amateur astronomer, has created in Mesklin an ultradense world of high gravity and equally high spin. The gravitational pull of the degenerate matter that accounts for most of Mesklin's mass gives the planet a polar gravity field about 665 times that of the Earth. However, Mesklin's rapid rate of rotation distorts the planet into the shape of an oblate spheroid (like a hamburger bun) and produces such an enormous outward directed centrifugal force at the equator that the gravitational field is almost totally cancelled there, leaving a net downward acceleration of just 3 gees. Of course, there are good grounds for worrying about the stability of such a world and for questioning the likelihood of finding degenerate matter in a planetary body, but *Mission of Gravity*, written long before the discovery of the rapidly rotating millisecond pulsars, is a wonderful example of sustained fictional invention underpinned by scientific rationalism.

Other examples of the use of science to supply exotic but credible settings for stories are *Dragon's Egg* (1980) by Robert L Forward, and Larry Niven's *Ringworld* which was mentioned in the last chapter. Unfortunately, Niven's marvellously conceived, if somewhat underused, alien artefact is mechanically unstable under small displacements with respect to the sun that it orbits. If the million mile wide, two hundred million mile diameter ring, with a mass of 2×10^{27} kg and an inner surface area approximately 3000 000 times that of the Earth, should ever drift away from its G_3 type central star, even slightly, there would not be any restoring force sufficient to bring it back to its equilibrium position. Happily, this problem, which arose out of an oversight on the part of the author, eventually led him to write a sequel, *Ringworld Engineers* (1980), in which the system used to maintain the relative positions of the Ringworld and its sun plays an important part.

Forward's book contains a similarly audacious act of speculation in its description of the Cheela, the tiny, and exceedingly flat inhabitants of a neutron star (Dragon's Egg) found to be approaching the Solar System from the direction of the constellation Draco (The Dragon). The idea that life might exist under the extremely rigorous conditions scientists predict on the surface of a neutron star—surface gravity 10^{11} gees, surface temperature 10^4 kelvin, and equatorial magnetic field strength 10^8 tesla—was first suggested by the astronomer Frank D Drake, whose name is associated with the equation used to estimate the number of intelligent, communicating species currently living in the Galaxy.

Some idea of the care and attention that Forward, a Senior Scientist with the Hughes Aircraft Corporation, has lavished on the creation of a scientifically credible background for his story may be obtained by studying the 22 page 'Technical Appendix' with which it concludes. The Appendix is presented as an extract from a twenty first century *Science Encyclopaedia*; it is a clever blend of fact, respectable scientific speculation, and pure invention. Where it assigns particular values to the physical properties of Dragon's Egg, those values are generally in good agreement with the generic properties of neutron stars listed in standard scientific sources such as Max Irvine's *Neutron Stars* (1978), or *Black Holes, White Dwarfs, and Neutron Stars: The Physics of Compact Objects* (1983) by

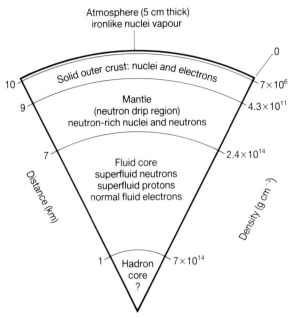

Dragon's Egg is a neutron star; the degenerate remnant of a star that has burnt out its nuclear fuel and collapsed under its own gravity. This cross section of a neutron star is taken from the appendix of Robert L Forward's *Dragon's Egg* but it agrees with similar diagrams in the conventional scientific literature.

Stuart L Shapiro and Saul A Teukolsky. It is the fact that Forward has obviously gone to some trouble to get the science right that makes *Dragon's Egg*, and its sequel *Starquake* (1985), such good examples of this kind of hard science fiction.

A final example of a science-based imaginary environment that deserves mention is another of Larry Niven's inventions; it is the 'smoke ring' of his novels *The Integral Trees* (1985) and *The Smoke Ring* (1987). Niven has explained that he derived the idea of the smoke ring from the findings of the spaceprobe Voyager 1 concerning the distribution of ionized atoms in the magnetosphere of Jupiter. (An atom is said to be *ionized* when it has lost one or more of its full complement of electrons; the *magnetosphere* of a planet is the region of space around the planet in which the planet's own magnetic field is the dominant magnetic influence.) It was known as long ago as 1974 that Jupiter's volcanically active satellite Io was accompanied by a banana shaped cloud of sodium atoms stretching for tens of thousands of kilometres along the satellite's orbit. But it was only with the arrival of Voyager 1, in 1979, that the full extent of the interaction between Jupiter's magnetic field and the particles liberated from the surface of Io became clear. In particular, Voyager revealed the presence of extensive clouds of sulphur and oxygen ions throughout the magnetosphere, with the low-energy ones concentrated in a torus (a doughnut shaped ring) roughly centred on Io's orbit.

Niven took this idea to the stars. He imagined a neutron star, which he called Voy, orbited by a giant gaseous planet rich in oxygen. He named the planet Gold, and gave it two large moons that could gravitationally strip away some of the planet's oxygen to produce a gas torus with an oxygen-rich centre surrounding the neutron star, somewhat similar to the sulphur torus surrounding Jupiter. He further imagined that the neutron star was part of a binary system and that the other component was a yellow dwarf star that could supply light and heat to the gas torus. The result of all this was the smoke ring; the oxygen-rich heart of the gas torus where the temperature and gas pressure are such that a number of different life forms can survive and indeed thrive. Living in a state of free fall, the various inhabitants spend their lives orbiting the neutron star. The smoke ring contains spherical ponds of water held together by their own surface tension, giant trees shaped like integral signs, various forms of indigenous flying life, and a number of human beings—the descendants of shipwrecked interstellar explorers.

(3)   *Using a piece of scientific information as the basis of a puzzle.*

This is science fiction's answer to the whodunit. Such stories could be called what-done-its. Isaac Asimov has written a number of stories of the kind featuring 'extraterrologist' Dr Wendall Urth, who uses his scientific knowledge and powers of deduction to solve interplanetary mysteries without leaving the confines of his university office. Another memorable story with a detective motif is British author Bob Shaw's 'The Giaconda Caper' (1976). Written in a Chandleresque style (Raymond, not A Bertram) the tale concerns a psychic detective hired to investigate the provenance of an almost perfect copy of Leonardo da Vinci's *Mona Lisa*. The picture resembles the original in every way except for a small separation of the hands, which are cupped together in the original. The investigator is sure that the picture is the work of da Vinci himself, but why should such a painting have been produced? Eventually the detective is led to a cave containing a centuries-old mechanical device and a large cache of Giacondas, all of them apparently the work of the master, all of them slightly different. With this discovery the scientific—or rather technological—point of the story becomes clear. The famous *Mona Lisa* is just one of the panels da Vinci produced for the world's first *zoetrope*—a rotating vertical drum with slits cut in its sides and a set of closely related pictures opposite the slits, on the inner surface of the drum. When the drum is rotated about its vertical axis, and the interior pictures viewed through the slits, the result is a moving picture, like a child's flick-book cartoon. Da Vinci was not only the inventor of the helicopter and the parachute but also, it seems, the father of the movies.

Not all scientific puzzle stories have an overtly detective theme. The most famous puzzle story of recent times is almost certainly 'Neutron Star', another work from the pen of Larry Niven, which won the Hugo award for best short story of 1966. (The first astronomical observation of a neutron star, by radio astronomers at Cambridge, was not made until 1967, but neutron stars had been present in the theoretical literature since their possible existence was first suggested by L D Landau in 1932.) 'Neutron Star' is part of the *Tales of Known Space* sequence, a

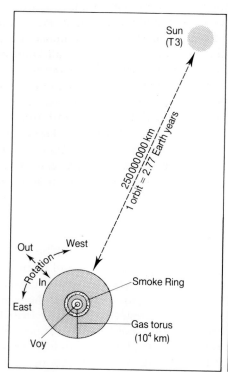

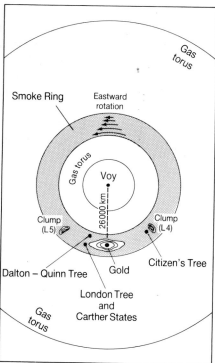

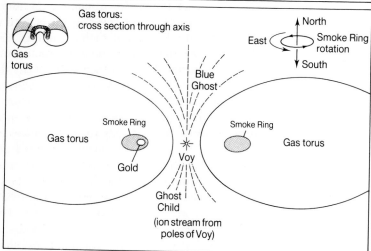

The Smoke Ring. Adapted from Shelley Shapiro's illustrations for Larry Niven's *The Integral Trees*. (Reproduced by permission of Larry Niven.)

loosely connected group of thirty or so stories that also includes Ringworld and spans more than ten centuries of galactic exploration and colonization.

The puzzle in this particular story concerns the death of two scientists who were investigating a neutron star. Using an indestructible spaceship supplied by a species of alien traders, the scientists should have been protected from all kinds of electromagnetic influence, yet, when their ship emerged from a close free-fall flyby of the neutron star they were dead, their bodies crushed to a bloody pulp in the nose cone of the spaceship. What caused their deaths? The alien species, anxious that their reputation as honest interstellar traders should not be damaged, employ Beowulf Shaeffer, the hero of several of the *Tales of Known Space*, to repeat the flyby and find out what went wrong. As Shaeffer approaches the star on a pre-determined hyperbolic trajectory he finds that his ship *Skydiver* shows an unexpected tendency to keep its nose pointed towards the neutron star. Shaeffer, located in the pilot's chair near the front of the ship, tries to correct the situation, but without success:

> I tried the maneuver again, and again the *Skydiver* fought back. But this time there was something else, something was pulling at me.
> So I unfastened my safety net—and fell headfirst into the nose.
> The pull was light, about a tenth of a gee. It felt more like sinking through honey than falling. I climbed back into my chair, tied myself in with the net, now hanging face down . . .

What is happening? Why is Shaeffer accelerating relative to the body of his ship? His first thought is that some sort of force is acting on him but not on his ship. He can think of no such force. He dismisses the idea in favour of the notion that something is pushing on the ship, deflecting it from its free-fall trajectory. But there is no sign of any such deflection from the predetermined course. As the ship heads towards the point of closest approach the force keeping the nose pointed towards the star gets stronger. So does the force pulling Shaeffer towards the nose. His life is in danger. Eventually, just in time to take the one course of action that can save him, Shaeffer realises the cause of the noseward pull—it is the tide.

The centre of mass of the ship is still following its preset trajectory, but the ends of the ship and all points in between are at differing distances from the neutron star and are therefore subjected to gravitational pulls of differing strengths. Each part of the ship has a tendency to follow its own independent trajectory, but is prevented from doing so by the rigidity of the alien material. Only Shaeffer, a 'free particle' within the ship, can respond to the tidal field which will inevitably draw him towards the nose of the spacecraft and crush him to death. He has one means of escape: to move further back, to the centre of mass of the ship where there is no tidal force. The alien traders, coming from a world without a moon, are unfamiliar with tidal effects; Shaeffer, accustomed to the influence of Earth's Moon, knows about tides and uses that knowledge to save his life.

Puzzle stories such as 'Neutron Star' are common in science fiction, though they rarely occur in such a pure form. Since they almost invariably involve a device or process of some kind they should perhaps be regarded as a subset of the following more general class.

(4)   *Using science to justify the existence of devices or processes.*

'Gadget' stories are a well recognized category of science fiction, although most of them do not involve much science. The majority of gadget stories rely upon imaginary science or scientific patter, but there are some which do not. Clearly, those devices described by Verne and Wells that have 'come true' must have had a reasonable scientific basis, though there is plenty of scope for argument about just how much foresight their predictions really required. Other devices still in the realm of fiction are almost certain to become realities. For instance, the statement that technical innovations will soon lead to the development of faster and more powerful computers stands an excellent chance of being correct; multi-billion dollar corporations are already hard at work making it come true. Such a prediction is little more than a modest extrapolation of present trends, more a piece of futurology than science fiction.

A device such as the space elevator in Arthur C Clarke's *The Fountains of Paradise* (1979) is a much more risky proposition. The idea of extending a cable from a satellite in a stationary geosynchronous orbit 36 000 kilometres above the equator to a fixed point on the surface of the Earth and then using it as a sort of railroad for lifting payloads into space has a fairly long history in the scientific and engineering literature. From the technical point of view the main problem seems to be the lack of a material of sufficient tensile strength to enable a cable of such immense length to bear its own weight. This sounds like the sort of problem that some future generation of materials technologists might well overcome. Even so, it's a very brave investor who will put all, or even some, of his hard-won savings into such a project at the present time. For the moment at least, Clarke's story still concerns a plausible gadget whose time has not yet come.

Some authors take a certain delight in constructing gadget stories which have a firm basis in science but which also contain some fundamental flaw that makes the gadget impossible. Clarke himself has produced a number of tales of this kind, which have been gathered together in his collection *Tales from the White Hart* (1957). They are often fun to read but it is debatable whether tales of osmosis bombs and phase-reversing sound eaters are examples of 'real' science in science fiction or simply elaborate pieces of scientific patter.

(5)   *Using the scientific process itself or using a credible scientific setting for a story.*

Good examples of this kind are rather rare. Many stories that involve scientists and scientific settings concentrate on the props of science or on conventional expectations rather than providing a balanced portrait of the real thing. In the older stories scientific research all too often involved a grey haired 'Prof', a female assistant in a white lab coat, a room full of explosive chemicals and a cage full of mutated rodents. In more recent stories it has become almost equally conventional that the 'Prof' must be female. Perhaps there is some justice in this development, but it still fails to reflect reality. As far as the scientific process is concerned the situation is even worse. The facts of science are widely used but the methods are hardly described at all. They are probably not sufficiently exciting.

Nonetheless, some very effective 'scientific' stories have been written, most of them by scientists. The one that comes most immediately to mind is Sir Fred Hoyle's *The Black Cloud* (1957), which was mentioned in the last chapter, but there are others. A story that has created a lot of interest in recent years is that told in Gregory Benford's award-winning novel *Timescape* (1980). Set in 1998, the book tells of a world going to wrack and ruin. Amongst those 'watching it come down' are two physicists working with faster-than-light tachyon particles in Cambridge University's Cavendish Laboratory. Against a background of declining funding, low morale and poor technical support, a background that will be all too familiar to many of today's researchers, the scientists stumble across a method of using tachyons to send signals back through time.

Tachyons have been respectable hypothetical particles since their existence was first suggested in the 1960s. However, numerous attempts to detect them have met with no success. Most researchers now take the view that they are simply mathematical artefacts that show up in certain quantum field theories, and that any theoretical indication of their existence as part of the real world is simply a sign that something is not quite right with the theory. The point about Benford's use of tachyons, as he himself has emphasized, is that *if* tachyons exist then science itself seems to justify their use in signalling across time. So that vital *if* of existence makes a good starting point for a serious science fiction story. Of course, since nobody knows how to produce tachyons, Benford has to make some use of scientific patter. He does so, but it is scientific patter of a very high order, scientific patter in the hands of a professional:

> "Well, we've got a large indium antimonide sample in there, see—" Renfrew pointed at the encased volume between the magnet poles. "We hit it with high energy ions. When the ions strike the indium they give off tachyons. It's a complex, very sensitive ion–nuclei reaction". He glanced at Peterson. "Tachyons are particles that travel faster than light, you know. On the other side—" he pointed around the magnets, leading Peterson to a long blue cylindrical tank that protruded ten metres away from the magnets "—we draw out the tachyons and focus them into a beam. They have a particular energy and spin, so they resonate only with indium nuclei in a strong magnetic field."

It's only fiction, but it would be a brave student who queried the correctness of such a description coming from a professor. Of course, descriptions of this kind, even such good ones, really belong to the earlier category of gadget stories. What is more significant about *Timescape* is the way in which Benford, a full professor of physics in California and a former visiting fellow at Cambridge, uses his insider's knowledge of physics to create an unusually realistic atmosphere. In many ways, *Timescape* is the 1980s counterpart to *The Black Cloud*.

(6)   *Using science peripherally, to justify a device or process, or to provide a generally 'scientific' background.*

This is the final catch-all category that includes all those stories that are clearly *science* fiction (as opposed to fantasy) but which really do not have very much science in them. Many of the works that fall into this class contain substantial amounts of imaginary science or pseudoscience, often interwoven with a little of the real thing. Prominent examples include some of the most popular science fiction stories ever written, Frank Herbert's *Dune* for example, and Isaac Asimov's *Foundation* series. Other examples are Gordon R Dickson's *Dorsai* stories, many of Robert Heinlein's books for the juvenile market, such as *Rocketship Galileo* (1947), *Time for the Stars* (1956), *Citizen of the Galaxy* (1957) and *Have Spacesuit—Will Travel* (1958), and David Brin's tales of 'uplift' as recounted in *Sundiver* (1980), *Startide Rising* (1985) and *The Uplift War* (1987). Although Brin's background is in physics, these last three are unusually rich in peripheral biology since they assume a billion year old galactic civilization in which new species are brought to an advanced stage of development by the genetic skills of older 'patron' races. The stories involve a wide range of aliens, many of them still indentured 'clients' repaying the patrons who uplifted them millennia ago. Only humanity, the first species for millions of years to achieve intelligence and starflight without being uplifted by patrons, refuses to play the game; declining to accept the indenture of the chimps and dolphins that had already been uplifted by human scientists before the first encounter with galactic civilization.

The clear lesson of the six classes listed above and the associated examples is that science enters science fiction in a variety of ways and plays a variety of roles. Sometimes it is in the foreground, being used to explain a piece of technology or to account for an unfamiliar process. More often it is in the background, justifying some particular aspect of either a real or an imaginary environment. Occasionally it may even provide the context for a story. Usually it is the facts of science that are of interest to the science fiction writer but, from time to time, some part of the scientific process may become the focus of attention. The science in science fiction comes in many forms and takes many guises.

Having considered the question of how science enters science fiction in some detail we now turn our attention to two other questions: How broad is the coverage of science?' and 'How accurate is it?'

It is very difficult to quantify a concept such as 'breadth of coverage' when dealing with science, and not particularly meaningful unless some idea of 'depth' is also supplied. In general terms however, few would deny that the coverage of science is broad but superficial. The superficiality is almost inevitable. There is an obvious need to avoid detailed mathematics and unnecessary references to phenomena with which readers are unlikely to be familiar. This practically removes all possibility of presenting proofs or demonstrations or even arguments other than those of the simplest kind. (Though some books, such as *The Black Cloud*, do include some mathematics!) Consequently, much of the science lacks depth and is not related to empirical evidence. It is simply factual information presented in descriptive terms and taken on trust. This is probably a great relief to the majority of readers and writers but in the wrong hands it can make science seem to be a body of established dogma rather than a process of investigation, argument and proof.

One source of the great breadth is readily apparent: the wide ranging backgrounds of the authors. Over the years science fiction has attracted writers with almost every conceivable kind of scientific interest, many of them technically trained. Indeed, there are a number of anthologies that contain nothing but stories by scientists; Fred Pohl's *The Expert Dreamers* (1963), for example, which includes contributions from Otto Frisch, George Gamow, Norbert Wiener and Leo Szilard, amongst others, and Groff Conklin's *Great Science Fiction by Scientists* (1962) with offerings from E T Bell, J B S Haldane, Julian Huxley and former Bell Labs Research Director John R Pierce, who made most of his contributions to science fiction magazines under the suitably atomic pseudonym of J J Coupling. Most authors have a natural tendency to make use of their own backgrounds when they write but this is perhaps especially strong in science fiction where displays of technical virtuosity are encouraged. An inevitable consequence has been the creation of a range of stories reflecting almost every field of scientific endeavour. Not only are the hard sciences such as physics, chemistry, biology and astronomy represented, but also the softer ones such as psychology, linguistics and anthropology. The stock of such stories is sufficiently great that a number of subject-based anthologies have been produced. Medicine, mathematics, psychology and anthropology have all been treated in this way, as have a number of other topics.

Naturally, the coverage of these subjects is far from even. Physics and astronomy have always been the most heavily represented, partly because of their intrinsic nature and partly because of the traditional themes of science fiction, particularly spaceflight. Biology, especially genetics, is also well represented though there is a tendency for the same few principles to be used again and again. Other fields, less able to provide snappy general principles or grandiose settings, often provide themes that run through a story. Ecology, for instance, permeates the whole fabric of Frank Herbert's *Dune*, yet a reader might not be able to state a single ecological principle upon completing the book's 510 pages. Psychology also provides the basis for a number of stories, although the extent to which its principles are made explicit varies widely.

In the case of physics the coverage is so rich and varied that it is possible to find stories that touch on almost all the major sub-fields of the subject. Some indication of this is given in table 2, though again the coverage is uneven. Special relativity is widely used, quantum mechanics hardly at all apart from a few general references and a widespread tendency to call on the Many Worlds interpretation of Everett, Wheeler and DeWitt to justify stories about alternate realities. A number of people

Table 2 Physics in science fiction.

Scientific topics	Fictional applications	Examples
Mechanics		
Newton's laws	Reaction drives	'The Martian Way' (Isaac Asimov)
Conservation laws	Teleportation	'Flash Crowd' (Larry Niven)
Centrifugal and coriolis forces	Artificial gravity in space ships	*Rendezvous with Rama* (Arthur C Clarke)
Orbits and ballistics	Interplanetary flight	*Space Cadet* (Robert A Heinlein)
Classical field theory		
Fields and potentials	Potential energy mountains	'What Goes Up' (Arthur C Clarke)
Electrostatics	Levitated moondust	'Dust Rag' (Hal Clement)
Electromagnetic induction	Power generation in space	'Tank Farm Dynamo' (David Brin)
Acoustics, optics and waves		
Interference	Sound 'eaters'	'Silence Please' (Arthur C Clarke)
Refraction	Slow glass for delayed light transmission	*Other Days, Other Eyes* (Bob Shaw)
Quantum theory		
The uncertainty principle	Justifying hyperspace!	*Mission to Universe* (Gordon R Dickson)
Interpretation	Many alternate universes	'All the Myriad Ways' (Larry Niven)
Quantum optics	Cooling lasers	*Sundiver* (David Brin)
Nuclei and elementary particles		
Nuclear energy	Atomic weapons	*The World Set Free* (H G Wells)
Tachyons	Communication with the past	*Timescape* (Gregory Benford)
Neutrinos	The harmless bomb	'The Neutrino Bomb' (Ralph Cooper)
Atoms and molecules		
The structure of matter	Travelling through solids	'The Fires Within' (Arthur C Clarke)
Properties of matter		
Superconductivity	Room-temperature superconductors	*Ringworld* (Larry Niven)
Thermodynamics		
Entropy	Second law reversal	*The Gods Themselves* (Isaac Asimov)
Maxwell's Demon	A warlock's air conditioning	'Unfinished Story No. 1' (Larry Niven)
Biophysics		
Bionics	$6 000 000 man	*Cyborg* (Martin Caidin)

Table 2 (continued)

Scientific topics	Fictional applications	Examples
Geophysics		
Internal structure of the Earth	Global catastrophe	*Don't Pick the Flowers* (D F Jones)
Astrophysics		
Neutron stars	Abodes of life	*Dragon's Egg* (Robert L Forward)
Black holes	Gateways to other universes	*Black Holes* (ed Jerry Pournelle)
Relativity and cosmology		
Special relativity	Twin paradox realizations	*Time For The Stars* (Robert A Heinlein)
General relativity	Time distortion	*Heechee Rendezvous* (Frederik Pohl)
Cosmology	Viewing the origin of the universe	*Tau Zero* (Poul Anderson)

have used this extensive body of popular literature as a resource in the teaching of physics. Papers describing such courses may be found in the *American Journal of Physics*, and Amit Goswami has written a book, somewhat curiously entitled *The Cosmic Dancers* (1983), that provides many good examples of the way in which discussions of basic principles can be based on science fiction.

In the case of astronomy the situation is somewhat different, but in many ways just as rich. Astronomical stories tend to deal with bodies rather than principles. There have been anthologies devoted to particular planets, to comets, to the Solar System as a whole, and inevitably to black holes. Not surprisingly, no one has seen fit to devote a collection to spectroscopy or polarimetry, but plenty of references to these and other techniques may be found. Subject-ordered references to several stories with an astronomical basis can be found in *Universe in the Classroom* (1985), the resource guide for astronomy teachers which Andrew Fraknoi prepared to complement William Kaufmann's well known college level astronomy text *Universe* (1985).

A general feeling for the range and depth of science in science fiction may be obtained from *The Science in Science Fiction* (1982) edited by Peter Nicholls. This lavishly illustrated 'coffee-table' book contains over eighty short but informative essays on topics ranging from stellar distances and nuclear fusion to cloning and biological warfare. The essays are arranged on a thematic basis, somewhat along the lines of table 1 (though there are many differences of detail and emphasis), and most of them contain references to specific works of science fiction.

The time has now come to turn to the question of scientific accuracy. To what extent do the writers of science fiction get the science right? In considering this topic it is important to bear in mind that there is no reason why the science in science fiction *should* be correct. If an author wishes to maintain that the world is flat, or the speed of light infinite, then there is surely no better context for such fantasies than the boundless universe of science fiction. Accuracy requires a considerable expenditure of time and effort on the part of an author; it creates many problems and brings few rewards. Little wonder then that many authors take the attitude that an unfortunate fact should not be allowed to stand in the way of a good story. Nonetheless, adherents to the school of 'hard-sf' do try to write within certain self-imposed restrictions that require at least a limited fidelity to the facts of science as they are currently understood, and it is primarily to their works that we should look for 'errors' in science.

Amongst the writers of hard science fiction there seems to be a general agreement that absolute scientific accuracy is not a realistic goal. Gregory Benford, writing on this topic in the magazine *New Scientist* (23/30 December 1976) has said that:

> Most writers think a reasonable standard would not fault a story unless the scientific or technical errors were visible to the lay reader— remembering, though, that the typical science fiction reader is relatively sophisticated in scientific matters and not easily fooled.

Of course, not all science fiction lives up to this standard. Most people can recall at least one example of a trivial error; light years being treated as units of time rather than distance, spacecraft making 'whooshing' noises in the vacuum of space, the finite speed of light being totally ignored or the word 'galaxy' being wrongly used to describe the Solar System. But such elementary errors, as we shall see in later chapters, are mainly confined to films and TV series; the errors that occur in books and magazines are usually rather more sophisticated.

Errors, whether visible to the lay reader or not, arise for a variety of reasons. One of the most obvious and pardonable sources of disagreement between fact and fiction is the progress of science itself. Mention has already been made of the way in which many scientifically based stories with Solar System settings, written in the 1950s or 1960s, have been rendered out of date by the discoveries of space probes such as the Pioneers, Mariners and Voyagers. Another source of error is the (perfectly understandable) failure of authors to examine their ideas in sufficient detail. Most science fiction stories could not withstand rigorous scientific analysis; it would be unrealistic to expect them to do so. Even when the basic idea of a story is plausible in its own right the author may fail to see its implications for other areas of science. This is a common problem since one of the characteristics that distinguishes the trained professional scientist from the scientifically literate layman is the professional's carefully developed 'feeling' for the articulation of science, his appreciation of the way in which one phenomenon impinges upon and influences another. An important part of a professional scientist's training is the development of a critical faculty for detecting ideas that do not fit in with reality, no matter how appealing they may be. It is this faculty that is often lacking in keen undergraduates and enthusiastic amateurs, and it is this same faculty that often seems to be absent from science fiction writers. The writers often have interesting ideas, but they do not ruthlessly reject those that are unrealistic—there is no reason why they should. There is little point in trying to present a catalogue of authors' mistakes but a few examples may be instructive at this point and others will be found in later chapters.

Teleportation is a favourite means of travel with many writers, from the 'Beam me up Scotty' of *Star Trek* to the 'Door' of Asimov's 'It's Such a Beautiful Day' (1954). Often such devices are simply unexplained spin-offs of imaginary science but some authors have tried to justify them by discussing the possibility of converting the matter of a body into energy and then transmitting that energy, along with a plan for its reconversion and reassembly, to some appropriate receiver. The amount of energy in the beam, and the quantity of information required to describe a living body with a full set of memories, makes such a device extremely implausible to say the least, but at least it seems to satisfy the conservation of energy. It doesn't assume something for nothing, or does it? In his essay 'The Theory and Practice of Teleportation' (1969), Larry Niven points out that any teleportation system that is constrained by the conservation of energy may well have to conserve momentum (the tendency of a body to continue travelling in a straight line) and angular momentum (the tendency of a rotating body to go on turning). These two additional requirements are frequently overlooked, but they could have serious, not to say lethal, consequences for someone teleporting from the equator to the pole. Perhaps teleportation devices, if they are ever built, will have to be fitted with cannons and propellers that can be used to shed unwanted momentum and angular momentum. If so, the result might look something like the accompanying figure.

Another idea that is popular with science fiction writers, and much closer to current technology, is that of using centrifugal forces, similar to those experienced

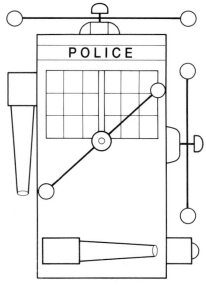

A teleporter.

when a car turns a corner at high speed, to provide a form of artificial gravity for space travellers. Those who have seen the film *2001* will probably recall the wheel-like space station that used this principle, or perhaps the huge drum onboard the *Discovery* that accommodated the astronauts during their flight to Jupiter. Science fiction authors are well aware that the inhabitants of such rotating habitats will not only experience a centrifugal force directed away from the axis of rotation but also a less familiar force due to the *Coriolis effect*. Unlike the centrifugal force, the Coriolis effect only influences objects that are in motion relative to the rotating environment. For example, the astronaut pictured standing on a rotating ladder in the figure would, as long as he remained stationary, only experience a downward centrifugal force. But as soon as he started to climb up the ladder he would also feel a force towards the left. If he climbed down the ladder he would feel a force to the right. These effects have been correctly described in a number of books, notably Clarke's *Rendezvous With Rama* (1973). However, the Coriolis effect is quite

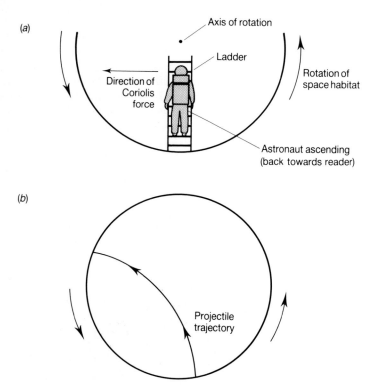

(*a*) Coriolis force in a rotating space habitat. The force only acts when the astronaut changes his distance from the axis of rotation. As the astronaut climbs the ladder there is a force to the left. If he were to descend, the force would act to the right. (*b*) The effect of Coriolis force on a projectile launched vertically upwards according to an observer moving with the rim of the habitat. The precise shape of the curve depends on the initial speed of the projectile.

complicated and confusion can easily arise. In the short story 'Small World' (1978) by Bob Shaw, for example, a projectile is described as travelling across a cylindrical space habitat along an S-shaped trajectory. In fact, the reversal of the Coriolis force after the projectile passes the midpoint of its course and starts its descent, means that the path is C-shaped when viewed from the drum, as shown in figure 5(*b*).

Despite the presence of 'errors' in many science fiction stories some authors take great pains to ensure that their speculations are as accurate as possible. Here is a quote from Bob Forward that shows the lengths to which he went when writing *Dragon's Egg*:

> For various reasons, I had the neutron star spinning at 5 revolutions per second. It turns out that "synchronous orbit" for an orbital period of 5 revolutions per second around a neutron star is at 400 kilometres, so I decided to put my human spacecraft in that orbit . . . I could have just assumed that I could do that, but my desire to make the science as accurate as possible led me to calculate the strength of the gravitational tides at 400 kilometres from a neutron star. It turned out they are 200 gees per metre! This is enough to literally tear you apart. I almost gave up at that point. I would have either to hope that the reader would not do the calculation that I did and go ahead with the story (Larry Niven did so in *Neutron Star* and got away with zooming a spaceship within a few kilometres of a neutron star), or give up and move the humans back where the tides were not so lethal. I then had an idea, and after a week of calculation proved to myself that six ultradense masses placed in a ring about the human spacecraft would make a counter tide that would cancel the tide of the neutron star. (I later turned these computer calculations into a scientific paper that has since been published in the *Physical Review. . . .*)
>
> [quoted from 'When the Science Writes the Fiction' in
> George Slusser and Eric S Rabin (eds),
> *Hard Science Fiction* (1986. Reproduced by permission of
> Southern Illinois University Press).]

Forward's paper, describing the way in which six 100-kilogramme tungsten masses may be used to cancel the Earth's tidal gravity field over a small volume within an orbiting satellite, was published in 1982 in the highly respected journal *Physical Review* D (volume 26 pp 735–44). Clearly, in the case of this particular proposal, the distinction between fact and fiction has little meaning.

Having surveyed the role of science in general terms it is now appropriate to take a detailed look at the way in which science fiction treats a particular scientific theme. The subject selected for this study, which occupies the whole of the next chapter, is time.

3

THE TIME FACTOR

One of the more intriguing parallels between science and science fiction is the fascination both have with time. The philosopher C E M Joad referred to 'the queerness of time' and it is time's strange quality that makes it play so important a role in science and science fiction, as well as in philosophy. From the beginnings of the twentieth century the mystery of the fourth dimension has been a subject to grasp, play with and interweave into the stuff that makes science fiction unique.

Of course, all fiction plays with time—condenses it, draws it out, refers back through reminiscence and projects forward in conjecture. Fiction explores many facets of time. The psychological perception of time may be presented as the stream of consciousness, concerned with the moment 'now', which has gone as soon as you are aware of it, or through the strange time of dreams. Mystical works explore the 'timeless' state while mythical tales make use of allegorical time. Time operates as a convention to be defined and re-interpreted in myriad ways. Very rarely does fiction work in what the computer industry calls 'real time', and cinema does so only in some exceptional cases, such as in Andy Warhol's one-shot films. Even actuality, newsreels and documentary films play around with time by editing—abbreviating events to fit a time slot. Television works in 'real time' rather more—live sport for example—but the conventions of drama, fiction and film assume an understanding that time is there to be played around with.

It is not this aspect of time, however, that is the subject of this study. The fourth dimension, time as a subject in its own right, enters science fiction just as it enters science, primarily through Einstein's theories of relativity, and the new concepts of space and time that these theories offer. With the advent of these theories time was no longer regarded as absolute; no universal clock ticked relentlessly on throughout the cosmos marking time everywhere in a perpetual, unchanging rhythm. Suddenly the possibility of strange time effects was vindicated by the most sophisticated of scientific theories, and the endless possibilities could be explored by the mathematical physicist and writer alike.

Time, in its psychological, fantastical and 'queer' aspects, can always be the basis for the unexpected in fiction, because the many different ways of perceiving it and trying to cope with it or understand it have always been there. Time is also a perennial problem in philosophy, and its 'queerness' has always impinged on literature. The past and future meet fleetingly in the *now* moment—which is gone as soon as it arrives, unless it is arrested into the timeless, mystical state. The change in scientific outlook on time, however, was still something unexpected and quite new.

Relativity theory is anticipated in fiction by H G Wells' *The Time Machine* (1895), where the notion of a fourth dimension is expounded by the Time Traveller to his friends at the start of the story.

> "There is no difference between Time and any of the three dimensions of Space except that our consciousness moves along it. A civilized man can go up against gravitation in a balloon, and why should he not hope that ultimately he may be able to stop or accelerate his drift along the Time Dimension, or even turn about and travel the other way?"

Wells' story was written ten years before the special theory of relativity, so it anticipates and heralds a more common access to the notion of time as another space-like dimension. Time as a dimension was not new to scientific thought, it was already inherent in dynamics and electromagnetic theory, but the media coverage of Einstein's ideas presented time to a general public in a fresh, scientifically exciting fashion. Time as the fourth dimension was perceived as quite mysterious by a public whose actual understanding of the concept of four-dimensional space–time was, and still is, at best hazy. Well's story, though, is not about relativity so much as time travel in a non-Einsteinian sense. We will examine it later. The most thorough and accurate relativistic exercise comes in Robert Heinlein's book *Time for the Stars* (1956) and it is also incorporated into the film *Planet of the Apes* (1968). Before exploring any stories, however, a brief account of Einstein's work is necessary.

The origin of Einstein's thought began with an attempt to reconcile the principle of relativity, that Galileo had introduced, with James Clerk Maxwell's equations of electromagnetism. In modern terms, the *principle of relativity* asserts that the laws of physics are the same for all observers—no matter how fast they move— provided they do not accelerate. In other words, a scientist working in a closed and isolated laboratory, with no view of the outside world, should be totally unable to determine the speed of the laboratory or even whether it was moving at all— provided it was not accelerating. Of course, in practice all laboratories on the Earth *do* accelerate, whether they are located in fixed buildings or moved around in trucks, trains or planes. Their acceleration is mainly the result of the Earth's rotation and its motion through space, but the acceleration of real laboratories does not in any way undermine the theoretical significance of the principle of relativity.

Despite its idealized nature, the principle of relativity, with its assertion of the essential equivalence of all unaccelerated observers, is of fundamental importance in modern physics. It rules out the possibility of defining an 'absolute standard of rest' or even of finding an 'absolute standard of reference', and it implies that the 'velocity' of a body only has meaning when related to some other (arbitrarily chosen) body. The statement that a car is moving at 60 mph, for example, only has meaning when we realize that we mean 60 mph *relative to the road*. It is in this sense—the sense embodied in the principle of relativity—that all unaccelerated motion may be said to be relative.

Einstein used the principle of relativity to guide his thinking. He took the view that any so-called 'laws' of physics that failed to accord with the principle of relativity must be in need of correction, or at least reinterpretation. One such set of 'laws' were those embodied in Maxwell's equations of the electromagnetic field. According to the then prevailing interpretation of Maxwell's equations, light (electromagnetic radiation) was expected to move through a universal medium called the ether at a constant speed, independent of the motion of its source. This particular interpretation of Maxwell's equations conflicted with the principle of relativity for if someone was travelling uniformly at the speed of light in a straight line and looking in a mirror at a light source behind them, then they would not be able to see that light and hence would know their absolute velocity. To resolve this conflict Einstein proposed that the speed of light, as well as being independent of the motion of the source, should also be independent of the motion of the observer. In other words, light cannot provide an absolute standard of reference. Hence his theory begins with the assumption that the speed of light is constant with respect to all unaccelerated observers, regardless of their relative motion. This assumption has never been falsified and actually makes sense of the famous Michelson–Morley experiment, which failed to detect the Earth's motion relative to light. (For a lengthier discussion of these matters see Shallis' *On Time* (1982).)

If light is measured to have the same speed with respect to any observer, regardless of relative motion, then something else has to give. Rather than changing Maxwell's equations Einstein reinterpreted time. As space and time are inseparable in Einstein's theory, space also has to have altered properties. Einstein's special theory of relativity describes these changes. It is called the special theory because it deals with the special case of a more general theory. In this special case Einstein is dealing with unaccelerated motion, i.e. uniform motion in a straight line, whereas the general theory of relativity allows for acceleration. In many applications the consequences of the two theories are essentially the same, but the general theory extends relativity to cover all possibilities.

In the special theory of relativity, when two observers are in uniform relative motion, each will find that time passes more slowly for the other. The relative motion that Einstein describes has the effect of slowing down time and of contracting space. In the general theory this effect ceases to be 'relative' if acceleration occurs; it then becomes 'absolute'. In an accelerating system time runs more slowly than for a system in uniform motion or at rest. By time here is meant the rate at which things change. The slowing down of time applies to all rates of change, for atoms, molecules or living cells. Clocks in an accelerating system run more slowly than when at rest, and this includes biological clocks. The accelerating observer also changes less rapidly, so time still feels normal, although it is passing at a different rate than for a non-accelerating subject.

In the general theory Einstein shows that accleration and gravity are equivalent and so time runs more slowly in a strong gravitational field than in a weak one.

Time ceases to be a uniform and universal quantity, as it was seen to be in Newton's cosmology and as it appears in ordinary life because it is a quantity which depends on both motion and location.

These seemingly fantastic ideas, introduced to resolve a logical anomaly between two older theories, turn out to be observationally and experimentally verifiable to a high degree, even though the effects only become dramatic at very high speeds and great accelerations. In the giant particle accelerators unstable particles exist for much longer when being accelerated around the circuits of the experiment than when they are at rest. Clocks at the top of a building run faster than those in the basement, where the gravitational field strength is greater, being closer to the centre of the Earth. The effect is very small and requires highly accurate atomic clocks to be detected, but when the experiment has been performed the predicted result has been found. Atoms in the Sun vibrate slightly more slowly than their equivalents on Earth, because of the greater gravitational pull exerted on them. The slower vibration is indicative of their time running slower than ours. There seems no end to the confirmation of these theories.

Both special and general relativity lead to the famous clock or twin paradox. This arises when one of two twins leaves Earth in a spaceship and travels away at high velocity into deep space. When he returns the travelling twin finds it hard to believe how old the stay-at-home twin has become, for time has passed much more quickly on Earth than in the speeding space vehicle. The 'paradox' lies in the apparent symmetry of the twins' observations which should imply equal ageing; its resolution consists in showing that accelerations render the observations asymmetric so that greater ageing of one twin is expected. Such a result has been confirmed experimentally by flying an atomic clock around the world. It registers less time passed, on its return, than the stay-at-home clock in the laboratory. Although the effect is miniscule at the speeds of jet airliners, the slowing down of time becomes increasingly dramatic the faster the velocity of travel becomes.

What cannot happen is acceleration to speeds greater than the speed of light, such as occurs in A E van Vogt's story 'The Storm' (1943). Hypothetical particles, named tachyons, as mentioned in the previous chapter, have the property of travelling only faster than light, with the speed of light as the lower limit of their attainable velocity. Just as particles we are familiar with require an infinite amount of energy to accelerate up to the speed of light so do tachyons require infinite energy to slow down to that speed. However, science fiction writers have been quite happy to make use of these imaginary particles in the cause of a good story, as, for example, in Bob Shaw's *The Palace of Eternity* (1969), where he postulates a tachyonic spaceship. If tachyons existed and could be observed by us they would appear to travel backwards in time and might be located in two places at the same time. Such notions appear to lead to a violation of causality and that is something Einstein's theory forbids, though physicists have speculated about loopholes that might allow tachyons to exist yet still preserve causality.

Conventional special relativity theory does not allow causality to be breached and implies that information can never be transmitted faster than the speed of light.

Neither does Einstein allow time travel in the sense of moving about in the time dimension as Wells allowed his Time Traveller to describe. All observers have their own local time but causal relationships must remain intact. The order of cause and effect cannot be reversed, even if notions of simultaneity become altered. We are each confined to our own time and only have access to times past determined by the passage of light, like observing a supernova star explosion that occurred thousands of years ago and whose light output has only just reached us.

The distortion of space and time described by Einstein's theories is seen at its most dramatic in black holes. Here gravitation has closed space in on itself and not even light can escape from its pull. Near a black hole time has slowed down to a virtual standstill and ceased completely at its boundary. Black holes are thought to exist in abundance throughout the universe as a logical end to the processes of the evolution of the more massive stars. A black hole is isolated from the rest of the universe by its 'event horizon', the boundary between its unfathomable interior and the space around it from which light can still escape. The slowing down of time in the vicinity of black holes has been used as an important story element in several science fiction tales. Especially notable is *Beyond the Blue Event Horizon* (1982) by Frederik Pohl, the second volume in his popular Heechee series. In this particular story entire alien civilizations are living more or less normal lives close to black holes while the rest of the universe, in which time is passing far more rapidly, continues to evolve.

In some scientific speculations the black hole might be interconnected to another part of the universe or even to universes in other dimensions, thereby allowing a traveller to reach those parts that would otherwise be inaccessible or even unimaginable. Such speculation, is, of course, just that—speculation—even though it may be mathematically quite respectable. Mathematics does not always have to map to the real world.

Even if black hole space–time warps allowed for such journeys, the principle of the non-violation of causality would remain intact. The journeys, even in theory, are not reversible. You may be able to travel through a black hole into another universe, but you cannot go back through the same hole to where you started. The reverse journey will always take you into yet another universe in a nightmare Kafkaesque fashion. Even when (and if) such space–time tunnels connect to another place in this universe travel down such a tunnel must still not violate causality. In science you cannot return to kill your father before you were conceived!

It should be noted that causality violations are not totally forbidden in the context of the general theory of relativity, though they may have some very strange consequences. Such violations and their possible applications were the subject of a 1988 scientific paper 'Wormholes, Time Machines, and the Weak Energy Condition' by Michael Morris *et al*, which created a good deal of popular interest at the time of its publication.

Some of these ideas from relativity are introduced into the *Planet of the Apes*. Here a spacecraft leaves Earth and travels out at a very high speed. It lands on a planet inhabited by superior, intelligent apes, who have outstripped human beings

Relativistic time dilation takes astronaut Charlton Heston into the Earth's future and a society where apes have evolved past mankind. (From the film *Planet of the Apes*.)

in an evolutionary sense, so that people are the ape's slaves and pets. The film deals with several interesting themes resulting from this scenario, while the hero, who is looking for a way to break free, find his spaceship and return home, suddenly discovers he is already back on Earth. He is a victim of time dilation and his initial journey was, in fact, a round trip to the future. The evidence is visually striking— the ruins of the Statue of Liberty protruding from the sea. The relativistic journey has enabled him to return home hundreds, or maybe thousands, of years later.

This is an excellent use of relativity theory in science fiction. The scientific idea has been accurately employed to allow us to explore interesting notions about evolutionary biology (albeit enhanced by the mutation effects of a holocaust), sociological complexities and, in the end, for a consideration of the nature of humanity. All this is wrapped up in an exciting story, with plenty of action. The sequels, however, push too hard on scientific credulity. Escape from the ape-ruled Earth happens, but it is an escape through time. Relativity is put into reverse to bring our hero back home not long after he originally left. Of course, he returns with some renegade apes, and the ape baby, brought up in American civilization, is the first link in the chain that leads to the apes becoming the dominant species. A long process of natural selection is not required here to bring about super-apes, they generate their own evolution from the future by returning to the past.

This breach of causality simply will not do. As a story-telling device it defies logic, although it sustains sufficient sequels to keep the popular and money-spinning series going. Relativity does not permit the return journey and so the consequences depart from the scientific realism that the original maintained so well. The social and humanitarian interests can be continued satisfactorily, but only by this weak link in the causal chain. The logical paradox of the extended story also defies credibility: you simply cannot have the result of a process initiating its origin. The baby ape in this story is its own forebear and science fiction becomes fiction leaving the science behind. It is perfectly legitimate in science fiction to play around with logical impossibilities by altering the rules, and an example will be given shortly, but the trouble with the sequels to *Planet of the Apes* is that, rather than stating a new set of rules, relativity theory is worked and misused.

Violation occurs in Disney's *The Black Hole* (1979), where an encounter with a black hole provides a topical backdrop for a space adventure. Many of the scientific prerequisites for the behaviour of objects in the vicinity of black holes are manifest, but not always correctly. The black hole itself, which would look distinctly non-visual on the screen, is made to glow red, whilst the escape from the doomed spacecraft, hurtling towards obscurity, would take far more energy than is presented on the screen. Several relativistic effects are ultimately attributable to energy considerations, including Einstein's equation that links matter and energy and which led to the development of nuclear power. Science fiction writers make use of the bizarre consequences of Einstein's theories often quite validly, as in Barry Malzberg's *Galaxies* (1975), where a whole world exists in a super-massive black hole, but more often they are simply used as a means of concocting and departing from the harness of science where narrative requires it. Not that there is anything wrong with that, but it is, in this type of analysis, important to realize where one is dealing with science and where with fiction.

The black hole as a time warp is simply a device, with a contemporary flavour to it, for transferring story material into weird environments, as, for example, in several *Dr Who* stories. Such a device allows an author to explore strange universes and altered environments, both physical and social. It pays lip service to current scientific understanding and hence places the material firmly in the science fiction arena. Very often the only distinguishing feature of a science fiction narrative may be that the authority of science and its concepts is used purely as a narrative device. What was excellent about *Planet of the Apes* was the ability to use scientific theory with accuracy. Having enough faith in a story to leave it open ended, leaving the consequences of the science borne out, as occurred in this case, can be seen as a courageous alternative to the classic 'happy ending', although it can be interpreted as a means to enable sequels to transpire.

The black hole, then, becomes one mechanism by which a traveller can cross unimagined tracts of space, to transport a story into another world, or to take the story into another time. Time travel is one of the most used devices in science fiction, allowing an author to explore events which, for many reasons, he wants enacted in an alien environment. Time travelling machines, beginning with that of the Time Traveller in H G Wells' story, are found in many guises, from the strange almost crystalline metallic transport of *The Time Machine*, through *Dr Who's* Tardis, caught in a time lock when it materialized as a police 'phone box, to the black hole. C S Lewis used a wardrobe in his celebrated book *The Lion, The Witch and the Wardrobe* (1950). It does not matter what the device is, the important thing is the space–time travel.

H G Wells' story begins with, and constantly refers back to, the time machine itself, although the machine is again simply a device allowing the author to present his own perspective on a possible future—one that includes a forecast of the two World Wars. The main thrust of the story lies in this form of forecasting and prediction and also in the social comment about the conflicts between different sorts of society. The main intent is not so much to explore questions of time, but to illustrate that different social bases appear in different guises and that their underlying form may not be readily apparent. This is social and political criticism, rather than science fiction, but we readily remember the science fiction element and the idea of time travel itself.

Time After Time (1979) does the same thing, linking H G Wells, Jack the Ripper and Sherlock Holmes in modern day California, thanks to a time machine. Here the device is used, not without humour, to erect a thriller with lots of popular ingredients. In *Dr Who* the Tardis provides the means of locating stories in any time period or any place in this or any other universe. The use the Tardis makes of the relativistic notion of space–time is deceptive. Its outside appearance bears no relation to its interior dimensions. You pass through the door from one world to another—a world that seems almost to lie outside normal space–time itself. The story, which has been developed for over twenty-five years, also has built into it the idea that the mechanism of the Tardis is faulty, enabling Dr Who and his companions to find themselves inadvertently in adventure after adventure. The controls are operated, the craft chugs and sighs and de-materializes from one place to another. Dr Who himself is a Time Lord, one of those elite who have been allowed to master time and space and for whom the universe is an unlimited panorama in which good battles with evil.

Woody Allen uses a crude and underspecified technology to transport him into the future in *Sleeper* (1973), so that he can exercise his satirical humour in a different setting. The Monty Python team do the same thing in Terry Gilliam's *Time Bandits* (1981). Much of science fiction is concerned with exploring ideas that would be difficult to set up in a contemporary setting—the exploration of other social orders, altered social and personal mores, the effect of different frameworks of thought and action, even the effect of different biologies or outcomes of evolutionary changes. These are all the concern of legitimate science fiction, and to explore them requires a different place or time. The idea of the time machine enables the protagonist to explore these other worlds and report back to us in our own time.

The technology for time travel is interesting in itself. As we do not have such a technology, or even an idea of how it could be built (indeed our science does not even allow for such a thing on theoretical grounds), the time machines themselves are fictional. H G Wells' time machine is apparently mechanical, but when designed for the film version of the story (1960) it combines what could be turn of the century mechanisms with lots of lights and the use, apparently, of electricity in a way that post-dates its setting (see the figure below). Dr Who's Tardis is much more attuned to what you would expect of the sixties and seventies. There is clearly lots of electronics around, lots of hidden power sources (maybe nuclear?) and a hint of things unknown to our thought. What else would you expect from a Time Lord?

In *Back to the Future* (1985), Spielberg's time-warp movie (he produced it and Michael Zemekis directed), the time machine takes the form of a car, filled with digital displays and a nuclear engine. Interestingly enough the car chosen was the DeLorean, both for its futuristic design and real but unusual aspect, and perhaps also to point up a joke about failure and controversy. To drive the car to another time you simply set up dates on a display board, very similar to the way it is illustrated in *The Time Machine*, although there the dials are very mechanical. The time travel is brought about by a Y-shaped set of tubes called a flux unit. Here, as in much other science fiction dealing with time travel, it is taken that time is contained in some sort of field, just as with gravity. Energy put into the field enables it to be

The Time Traveller, Rod Taylor, prepares to travel into the future on his time machine. (From the film of H G Wells' *The Time Machine*.)

traversed faster than normal, which sounds plausible, in that it uses scientific jargon, but is actually unconnected to any existing physics. Incidentally, the inventor of this time machine can't work out how to design a flux unit, but the youngster who has gone back to the past is able to tell him what it looks like (or will look like), hence the invention depends on it having been invented and that future knowledge is transmitted back in time. This is another illogical temporal loop. How time machines actually work is never really specified, except in terms of some pseudoscientific jargon. Space travel is far more accessible to represent technologically but with time we have to suspend disbelief and take what is given us as part of the narrative such science fiction offers.

The time warp also does not need to be specified. The idea that such discontinuities in time may exist is accepted as a convention in science fiction and is loosely based on ideas about black holes, white holes and tunnel effects through the space–time matrix. The discontinuities may be actual trap doors, holes or similar devices, as in *Time Bandits*. They may be generated, as in the warp factors of such basic stock as *Star Trek*, where the Starship Enterprise can gear up its warp drive to move through the science fiction convention of hyperspace. Alternatively the device for a time warp may exist in another dimension, which may be psychological or psychic, but with material consequences, as in the *Garth* science fiction strip cartoon. The main thing is to be able to travel to other places and other times. In Anderson's *Tau Zero* (1970) one wonders where the characters actually end up. Their spaceship travels so fast that relativistic time dilation slows down their local time to such an extent that they witness the recollapse of the universe back into the mirror image of the original big bang. As the universe contains the whole space–time matrix they must be left cruising about nowhere in space or time!

An unusual television series, *The Time Tunnel*, employed as its technology a manufactured time warp—a tunnel through space–time generated in the laboratory. It enabled the time travellers to turn up at many real historical disasters and do some good. However, the producers, not being able to distort history beyond recognition, ensured that causality was not violated. When, for example, the time travellers found themselves aboard the Titanic, an hour or two before the disaster struck, they were regarded as stowaways and locked up. The Captain refused to listen to their predictions of the impending catastrophe but a more junior officer took their word on trust. When the great ship hit the iceberg he had them released and their efforts, as well as their warnings, helped save lives. Had they not been there the disaster would have been even worse.

Causality is deliberately violated, however, in *Back to the Future*. The point of the story is to change the present into a more acceptable form by returning to the past in order to alter those things that resulted in the present that the story began with. In travelling back to the fifties the hero disrupts the events that 'originally' led to his parents meeting and falling in love. In order for the hero to be conceived he has to make amends for his being in the wrong time, but in doing so he alters his family's history in such a way that when he returns he finds his world is much better than when he set out. The illogicality is beautifully brought off, with much humour and a

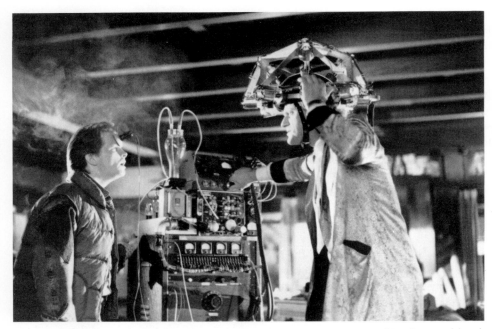

Time muddled up. Michael J Fox travels back to the 1950s in a time machine invented by the 'mad' professor, who is still struggling to perfect a time-travelling device. (From *Back to the Future.*)

delightful play on impossible logics. For example, the hero loves rock and roll, so when he goes back to the fifties and is asked to play in a band he plays his favourite music and invents rock and roll!

The logical impossibilities that time travel gives rise to is exploited quite brilliantly in Robert Heinlein's story 'All You Zombies' (1959). The narrator is a temporal agent, permitted to travel in time at will, using a time machine that conveniently packs into a suitcase. (It contains a net that encloses the time field that is manipulated to effect the time travel.) He tells us the story of a man who was born a girl, abandoned as a baby at an orphanage gate, who later grew up, got pregnant and gave birth to a baby daughter who was abducted from the hospital. The trauma of childbirth triggered an instability in her metabolism in such a way that she changed into a man—the unmarried mother. The temporal agent, there to recruit this man into the time travelling service, takes him into his past. There he meets himself when he was a woman and makes love to her (himself). The agent then removes the resulting baby girl and takes it back to the more distant past, depositing it at the orphanage gate. Thus this man/woman is his own mother and father. The final twist in the story comes when the agent recruits the man into the service and we discover that he has recruited himself.

The internal logic of this story makes sense only when you work backwards in time. The man recruits himself as a temporal agent, having made love to himself as a woman, with the resulting child growing up to be the woman who changes sex. Read forwards in time the story seems highly paradoxical, but on analysis this story of strange disruption of normal causality, resulting in a kind of immortality for the hero, which makes everyone else seem like zombies, works perfectly well when the future causes the past to happen. By changing normal rules of causality, by permitting time travel, Heinlein constructs a masterpiece of the genre.

The alteration of history is sometimes used in science fiction to induce the unexpected twist in the tale of the story, and the result is pure surrealism. Salvador Dali's paintings, at the time of high surrealism, invoke a real feeling for time through the image of the melted and draped clock placed on an alien landscape. This imagery owes more to dream psychology that it does to science fiction, but nevertheless stories such as *Time Bandits* or even *Back to the Future* can be seen as having surrealist roots. Distortion of time effects, which are by no means confined to science fiction, are very much the stuff of story telling. Einstein's theories of relativity could themselves quite legitimately be regarded as surrealist, as Dali was aware, and the strangeness of time warps and time dilation becomes more comprehensible when considered in relation to the 'queerness' time often has in dreams. Time tricks, however, crop up in science fiction in other forms as well.

Rather than travel about in time, another writers' ploy is to travel to a place that has become 'frozen' in time. *The Lost World* (1960), based on a Conan Doyle story, makes use of this device to enable the characters to discover what life was like at the time of the dinosaurs. A location is found whose isolation from the rest of the world is such that evolutionary changes have not occurred there. This leads to great fun with monsters of all sorts. The lost world of the title is an almost inaccessible plateau in South America, where nature remains as it was two million or more years ago. *The Land That Time Forgot* (1974), based on an Edgar Rice Burroughs story, reworks the same material.

Such encounters with the past spring from the possibilities opened up by the exploration of remote places in the nineteenth century. We are here walking in the territory of Jules Verne. At that time there was a different degree of credibility to such stories, whereas the late twentieth century can only handle them as escapist entertainment. For all that, encounters with pterodactyls, tyrannosaurus rex and the whole array of prehistoric animals makes for great material for legitimate science fiction. Such lost worlds enable the writer to genuinely deal with other realities and strange environments in a way that is based on scientific fact, and to provide an element of education alongside a great yarn.

In the greatest of all the prehistoric monster films, *King Kong* (1933), the discovery of a lost world is combined with a morality lesson on ecology, contrasting the civilized world with the purer but more primitive mores of nature. The destruction of King Kong shows our misunderstanding of what our relationship with nature should be—it is not there to be displayed, tamed and desecrated, even if it may also act as a metaphor for our primitive instincts. Such lessons in our attitudes to the natural world and the exploitative propensity of modern man come wrapped up in the exploration of the land forgotten by time, which enables the special effects man to grapple with the rigours of realistic model making.

The past in the present. In the film *The Land that Time Forgot* a pterodactyl makes off with one of the explorers of a lost world.

Time slips are another device used to transport people to different eras. The idea that times can get muddled up has been exploited in several ways. Gordon Dickson's novel *Time Storm* (1977) is based on the idea that the time matrix has suffered a violent convulsion so that different places on Earth are displaced into different historical eras. This allows a caveman, a twentieth century hero and an alien visitor from the future to engage in mutual relationship. Fred Hoyle used this same idea in *October the First is Too Late* (1966), where different European countries are fixed at different historical times. If you want to visit 425 BC then go to Greece.

At the end of Wells' *The Time Machine*, when the Time Traveller has departed for the last time, the narrator speculates as to whether he has gone back to the future or slipped away into the past, to roam the Earth among the dinosaurs. In the late twentieth century, of course, we have to travel to other worlds to encounter such creatures. Instead of finding lost worlds on the Earth we have to discover them on distant planets. *Star Wars* and its sequels made imaginative use of such conventions, producing a wide range of creatures to fascinate us.

The alternative is, of course, to invoke some mishap or catastrophe which produces large monsters from otherwise everyday creatures. So we have been

Vision of the future. Harrison Ford inhabits the bizarre contrasts of style to be found in the world of *Blade Runner.*

entertained by giant ants, wasps, spiders and so on. Changes of scale, as with changes of time, provide the science fiction writer with another device for presenting us with a fresh perspective on ourselves and our world. It is a tradition that extends at least as far back as Jonathan Swift and *Gulliver's Travels*, and probably, as indicated in Chapter 1, even further.

While these stories take us back to a past found in the present, we find ourselves in the future in most science fiction. Whether exploring the future has much to do with time and time travel is another matter, but a basic stock in science fiction is to explore a future world. Without invoking time machines the science fiction story of the future itself transports us to a time yet to come. Here we can explore the advancements made in civilization, or the regressions, or become acquainted with new technologies, fashions and societies. The future may be an excuse to rework the themes of another genre, as in *Blade Runner* (1982), where crime detection is set in a future that has been imaginatively portrayed, or *Westworld* (1973) which reworks the western theme. In the great, classic film, *Things to Come* (1936), scripted by H G Wells himself, the future is explored in terms of politics and social change, suggesting that there is and should be a limit to scientific progress.

Works like Asimov's *Foundation* trilogy not only draw up a long and complicated future society, spreading across the galaxy, but also include their own history, some dating back to our own time. An alternative to the future is to transport the audience to another place, so that location in time is irrelevant. In *Star Wars* we are told that the story takes place 'a long time ago in a galaxy far away'. As well as removing us from the constraints of what we know, the 'Once upon a time . . .' opening tells us that we are to be presented with a fairy tale, in classic mould, even whilst dressed in the costumes and convention of science fiction.

The common sense of illogic that at first appears to lie in Einstein's theories play no real part in such sagas. What they have done, however, is to make a framework of time, that seemed so secure, linear and sensible in Newton's universe, dissolve in our minds and turn time into a dimension which could be explored with a legitimacy that did not exist before our scientific precepts were altered. Wells' attunement to the time, his writer's antenna, gave him an understanding of what was happening in science, even before the theory of relativity unlocked the door to the probable pasts and possible futures which have become such an important feature in twentieth century science fiction. What was available to Swift and Verne was spatial exploration, with changes in scale or the discovery of time anomalies in the real time of their day. Wells changed all that when his Time Traveller discovered that final factor—the fourth dimension—time.

4

DOMESTICATING SCIENCE

The year is 1951. The place, Washington, DC. The war still rages in Korea, 6000 miles to the East; closer to home, Ethel and Julius Rosenberg, the atom spies, languish in prison, awaiting execution. As the witch hunts gather momentum, Americans become suspicious of outsiders and of one another. Right fears Left and Left fears Right, while those in the centre fear contamination from both sides. The immigrant, the dissenter, the intellectual—anyone whose face doesn't fit or whose opinions don't conform—is suspect.

Yet, on this clear, balmy day in the American capital it is easy to put aside thoughts of conflict and division. Children play on the well-manicured lawns, dogs take their owners for a walk, and people laze in the sun. The world is at peace with itself.

All of a sudden, a black speck appears in the sky. It grows. Is it a bird? A baseball? Perhaps a Soviet buzz-bomb? No—it is a flying saucer. As the silvery disc is sighted, pandemonium breaks out. Newspapermen scuttle to and fro, police cars tear down the streets, and tanks gear up for action. The President declares a national emergency. As the saucer glides down to Washington Mall, the military prepares to surround it—howitzers, machine-guns, cannons move in. The shining saucer lands, and from its seamless surface glides a ramp, down which walks Klaatu, a tall, handsome humanoid, accompanied by a huge metallic robot, Gort. A rustle of panic moves through the crowd that now encircles the alien visitors, but Klaatu offers a reassuring smile and holds up his right hand in the universal gesture of greeting. 'We have come to visit you in peace and with goodwill', he says in modulated English, as he reaches into his pocket to draw out a gift. Sensing danger, a nervous soldier draws his pistol and fells him with a single shot.

Fact or fiction? The answer is a bit of both. The scene opens the film *The Day the Earth Stood Still* (1951), based on the story 'Farewell to the Master' by Harry Bates (1940), but it so powerfully reflected contemporary fears and aspirations that many cinema-goers would have found its documentary approach and topicality striking; indeed, the movie's producer Robert Wise was accused of planting reports of UFOs in the press to stir interest in his film. The fifties witnessed the blurring of boundaries between science and pseudoscience, and the emergence between the two of a deep fascination in matters concerned with flying saucers, aliens, robots and UFOs.

Although the finer details of the plot and characterization may have been forgotten, those who have seen *The Day the Earth Stood Still* will remember the powerful, even intimidating, soliloquy with which Klaatu bids farewell to Earth. We have learned that a distant planetary federation, disapproving of the Earthlings' policy of atom-bomb testing, has dispatched Klaatu on a diplomatic mission to warn governments locked in a cycle of war and aggression to come to terms with their differences or face being blown apart. He has come in the name of cosmic peace and is the victim of attack and abuse, hypocrisy and fear, misunderstanding and hostility. The leaders of nations across the globe have put petty rivalries before world harmony and military authorities have conspired to undermine his mission. Only the community of scientists, along with symbolic representatives of innocence (a child) and common sense (a woman), have shown appreciation for the urgency and merit of Klaatu's case. So it is with a tone of regret, tinged with ill-feeling, that he takes his leave:

> The Universe grows smaller every day—and then threat of aggression by any group—anywhere—can no longer be tolerated. There must be security for *all*—or none is secure . . . This does not mean giving up any freedom except the freedom to act irresponsibly. Your ancestors knew this when they made laws to govern themselves—and hired policemen to enforce them. We of the other planets have long accepted this principle. We have an organisation for the mutual protection of all planets—and for the complete elimination of aggression. A sort of United Nations on the Planetary level . . . The test of any such higher authority, of course, is the police force that supports it. For *our* policemen, we created a race of robots. Their function is to patrol the planets—in space ships like this one—and preserve the peace. In matters of aggression we have given them absolute power over us. At the first sign of violence they act *automatically* against the aggressor. And the penalty for provoking their action is too terrible to risk.
>
> The result is that we live in peace, without arms or armies, secure in the knowledge that we are free from aggression and war—free to pursue more profitable enterprises.
>
> We do not pretend to have achieved perfection—but we do have a system—and it *works*. I came here to give you the facts. It is no concern of ours how you run your own planet—but if you threaten to *extend* your violence, this Earth of yours will be reduced to a burned-out cinder.
>
> Your choice is simple. Join us and live in peace. Or pursue your present course—and face obliteration. We will be waiting for your answer. The decision rests with *you*.

> [Quoted from Seve Rubin, 'Retrospect', *Cinefantastique*, 4 (1976)]

The Day the Earth Stood Still was not the first film of the post-war period to take up, distil, and reflect contemporary interests. Indeed, Darryl Zanuck at 20th Century Fox decided to go ahead with the project partly because of the recent box-office success of George Pal's *Destination Moon* (1950), which was doing unprecedented business across the United States. He was also well aware that other studios were investing in the genre and was looking forward to cashing in on the new craze: Howard Hawks was preparing *The Thing* (1951) at RKO, Monogram was busy making *Flight to Mars* (1951), and Pal was already at work on a follow-up science fiction project to *Destination Moon*.

In this chapter we shall be looking in some detail at the *The Day the Earth Stood Still*, paying particular attention to the contexts in which this and similar films were produced and screened, and considering the impression science fiction films of the fifties gave of the nature of science and its place in society. Then, in the next chapter, we shall turn to Hollywood's treatment, or rather its manufacture, of the figure of the scientist as 'regular guy'. As we shall see, Hollywood's science fiction version of the scientist was remarkably subtle and coherent and owed little to the earlier stereotypes—the eccentric boffin and mad doctor. The reason for this, in part, was that it incorporated, and reflected, some of the contemporary burgeoning interest in science and technology. It also took account of the political and social situation of fifties America, and the needs of the film industry itself.

America emerged from the war a major nation on the world map, committed now as never before to interventionism across the globe, and with a sense of mission, energy and exhilaration at home. Material well-being coupled with national pride bred a sense of power and complacency, deriving as much as anything from a general lassitude toward politics and foreign wars. Bursting with the enforced savings of the war, Americans were avid consumers after it, and the GI Bill of Rights (which became law in 1944) helped returning veterans to borrow money to set up businesses or to attend college. A baby boom signalled the feeling of ease and sense of future promise that gripped the country.

On the foreign scene America experienced a mixture of tension and fear, resulting from its rather sudden immersion in international affairs. Tensions between the US and USSR increased when Stalin, in quest of greater security, broke the Yalta agreements. Despite the dream of an international pact embodied in the UN, the Cold War, as it came to be known, escalated, and relations between the US and USSR became tainted with the suspicion and distrust which continued even after the situation in Eastern Europe stabilized in 1949. Indeed, the fifties witnessed a distrust of communists and leftists which grew into fear and persecution. A list of spy-ring revelations, the explosion of a Soviet A-bomb and the fall of mainland China to the communists fanned these fears into hysteria.

In the vanguard of this hysteria was the now infamous activity of the House Un-American Activities Committee, which had lain dormant during the war years, when the US was allied to the USSR. It began again in earnest in 1947 when it put the activities of the motion picture industry under the spotlight. Though the decade opened with American entanglement in the Korean war and witnessed the increasingly repressive and paranoid atmosphere generated by the investigations of Senator McCarthy, a climate of conformity was sustained at home. America was steered by General Dwight Eisenhower, its President from 1952 to 1960, through domestic crises and intervention abroad, with remarkably little disruption of the domestic compact. The American may well have felt like a sailor abandoned on a desert island, firmly implanted on Earth, but surrounded by stormy seas.

The atmosphere was a curious mixture of tension and self-confidence, of enforcement and consensus; and the ability of Hollywood to sustain and advance this precarious state of affairs was, in retrospect, remarkable. Films watched by rich and poor alike, in country and city, created a powerful illusion of social cohesion and stable values. Ingredients of romance, action and social comment were combined in a popular, often extravagant recipe. Cinema culture—'mass culture', to use the idiom of the day—was democratic, at least in the sense that it refused on the whole to discriminate against anything or anyone. Of course, there were limits to its tolerance, but the manner in which movies could offer a comfortable mix, a scramble of homogenized culture, was part of their power. This was a process, as critic Dwight Macdonald noted at the time, which 'destroys all values, since value judgements require discrimination, an ugly word in liberal–democratic America' (quoted in Brookeman (1984)). This conformity was by no means disrupted by intellectual culture—rather the reverse, as many fell in with the signal of Daniel Bell's collection of essays *The End of Ideology* (1960) in favour of pluralism and the perfectability of US values and society. Stability, not to say complacency, were the watchwords, even though nuclear armageddon hung over the decade.

In many ways, the fifties represented a watershed in the development of cinema, for they saw a dramatic decline in audience levels. After the highpoint of 1946, when the average weekly cinema audience in the US was 66 million (in Britain the figure was 31 million), the figure dropped to 50 million (26 million in Britain) in 1950. Even so, it is difficult to overestimate the importance of cinema: in Britain, for example, there were over 1300 million admissions in 1950, which was ten times the combined total for football matches, theatres and music-halls (figures from Browning and Sorrell (1954)).

But television was Hollywood's menace, increasingly so after 1951, when for the first time it linked the whole of the US; in Britain, attendances slumped a few years later once television sets became available to working class families and, perhaps not coincidentally, once commercial television came into being. The cinema responded by offering what television could not: size, spectacle and, through a variety of technological innovations (the triple-screen process, 3D, CinemaScope), overwhelming personal involvement. Jack Warner forbade the appearance of a television set in a single frame of any Warner Bros picture, but this could hardly delay the irrevocable shift in the pattern of popular entertainment.

By 1959 (the year which saw David Niven kicking a set in the television-breaks-up-a-family movie *Happy Anniversary*, and 20 years after *S.O.S. Tidal Wave* (1939), probably the first ever media corruption movie) the average weekly attendance in cinemas had fallen to 22 million, and in Britain to half that figure. The warning offered earlier in a ludicrous science fiction film, *The Twonky* (1953), in which a demonic television takes over the life of philosophy professor Kerry West, doing all the household chores and mesmerizing those who stand in its way, had plainly gone unheeded. Though audiences still went to their local cinemas in great numbers, they could also find entertainment quite safely at home.

Hollywood's attempts to breathe new life into motion pictures resulted in an unprecedented interest in science fiction films. This coincided with, and did much to nourish, an explosion of interest in matters scientific, particularly in space travel. In 1950, for example, the Hayden Planetarium announced that it was open for bookings for voyages to the Moon. Within a few days, 25 000 people from around the world hastened to apply, gripped by the fever for space travel, which was spread by anything from pulp magazines to the prognostications of experts like Werner von Braun.

The fifties saw the prestige of science and technology in the United States mushroom like an atomic cloud. In schools and universities, more students than ever before were choosing some branch of science for their careers. Military budgets were being earmarked for scientific research on an unprecedented scale. Books and magazines devoted to science were rolling off the presses in greater numbers than at any previous time in history. Even in the realm of popular literature, science fiction was threatening to replace the detective story as the most popular genre. And closely connected to these developments, the fifties saw a boom in the number of esoteric, purportedly 'scientific' theories, riding on established, reputable research. The decade not only witnessed the first wave of mass reports of flying saucers and other UFOs, but also the spectacle of thousands of ordinary folk entering for the first time 'dianetic reveries', sitting expectantly in 'orgone boxes', or having their eyes opened by such bestsellers as Immanuel Velikovsky's *Worlds in Collision* (1950). Food fads, medical cults and religious sects all cashed in on—and helped to nurture—an explosion of public interest in genuinely or purportedly scientific matters.

The film industry's interest in science fiction was also a response to popular features on radio and television. Before Hollywood ever got airborne, the comic strips, radio and cereal packets had already filled the interstellar regions with traffic as thick as in rush-hour Manhattan. Fantasy thrillers of various sorts had been common on US radio for the previous two decades (a highlight had been Orson Welles' famous production of 'War of the Worlds' on CBS in 1938), but science fiction came into its own in 1949 when an anthology of recent work, 'Dimension X', later retitled 'X Minus 1' was broadcast on NBC. In the UK meanwhile, BBC radio for some years broadcast dramatizations, single plays, serials and readings by distinguished authors such as H G Wells, Isaac Asimov, John Wyndham and Arthur C Clarke, and continued through the fifties with popular serials like 'Journey into Space', a transmission of which in 1955 reached 5 million listeners.

The first science fiction series to appear on American television was 'Captain Video', aimed primarily at children, which began in 1949, and this spawned a host of imitations such as 'Buck Rogers' (1950), 'Tom Corbett, Space Cadet' (1950), 'Superman' (1953) and 'Space Patrol' (1954). 'Out of this World', which started in 1952, was a more serious approach which mixed science fiction with science fact and regularly featured guest speakers interrupting the story to explain scientific principles, an approach continued in the 1955 series 'Science Fiction Theater'. In Britain, 1949 also saw the first televized science fiction event—George Orwell's '1984' (published in the same year)—followed five years later by the BBC serial 'The

Quatermass Experiment'. It is an indication of the close, not to say parasitical, relationship between the different media that such programmes, if successful, were rapidly put onto the big screen; after a second Quatermass serial in 1955, two featured films followed, *The Quatermass Xperiment* (1955) and *Quatermass II* (1957). American audiences had not benefited from the British television productions, so the titles were changed to the more easily assimilable *Enemy from Space* and *The Creeping Unknown*, respectively.

Hollywood has on more than one occasion had to spend its way out of trouble, but to out-Keynes Keynes in the fifties took substantial investment, and there was at first a good deal of reluctance in attending to the great public interest in science by investing in suitable movies; perhaps the memory of spectacular financial disasters like *Metropolis* (1926) and *Just Imagine* (1930) weighed heavily. In his 1950 article 'Shooting Destination Moon' Robert A Heinlein claimed after much time and effort was expended in obtaining backing for *Destination Moon* that 'the money men in Hollywood write large checks only when competition leaves them no alternative; they prefer to write small checks, or no checks at all'.

Once committed to their various projects—and inspired by the profit motive rather than any enlightened desire to enhance public appreciation of science and technology—film financiers moved into science fiction with speed and relish. The result was a series of movies of enormous diversity of theme, quality, and production, with the most memorable being such features as *Destination Moon* and *The Day the Earth Stood Still*, as well as *When Worlds Collide* (1951), *War of the Worlds* (1952), *The Creature from the Black Lagoon* (1954), *Them!* (1954). *This Island Earth* (1954), *Forbidden Planet* (1956), *Invasion of the Body Snatchers* (1956), *The Curse of Frankenstein* (1957), *The Incredible Shrinking Man* (1957), *I Married a Monster from Outer Space* (1958) and *The Time Machine* (1960).

If these represent the riches available, what other pleasures could audiences expect to share in their warm, crowded and ornate picture palaces? The answer is an enormous variety—the good, the bad, and the camp. Indeed, some of history's 'best-worst' movies have crawled out of the swamp of fifties science fiction. Movies such as *The Robot Monster* (1953), *The Green Slime* (1954) and *From Hell it Came* (1957) have recently been rediscovered by film buffs, pulled out of their archives, dusted down and screened in film seasons devoted to kitsch, bad taste and tackiness. Viewers are then confronted with what are the true gargoyles on the architecture of the cinema, to be looked at with a mixture of awe and pity. Such films—and there are plenty more fish of that quality in the cinematic sea—make the grade of truly awful, offering fine examples of wooden acting, lame scripts and rickety props. They have turned epithets such as 'pulp', 'sleazy' and 'trashy' on their heads, making them terms of approbation. But their quality says nothing about their historical importance, and though we shall not be focusing on them in the remarks that follow, we do not consign them peremptorily to history's dustbin. Rubbish is rubbish, but the history of rubbish can throw light on the underbelly of culture.

A decade that can produce the perverse delight of *The Green Slime* and *Forbidden*

A rescue party arrives on a desert island to find that the scientists they have come to pick up have disappeared and the island is slowly sinking into the sea, as a result of the tunnelling beneath it of enormous mutant crabs. *Attack of the Crab Monsters* was a fine 'good-bad' movie and one of Roger Corman's best—lively, direct and almost believable.

Planet clearly presents us with a broad and multicoloured spectrum of films from which to choose; indeed, it is difficult to conceive of any other genre which encompassed such a range. There were black and white, colour, technicolour, CinemaScope, and 3D science fiction movies. There were serials and feature films, A-movies and B-movies. There were lavish studio productions alongside movies crippled by low budgets. *Destination Moon* cost just over $600 000 to make; its rival, *Rocketship XM* (1950), was produced for a mere $94 000, shot in three weeks, and to save money depicted a landing on Mars rather than the Moon, thus allowing its producer to make use of real desert scenery rather than having to build moon sets within a sound-stage. (Even special effects wizards like Ray Harryhausen had to work to tight budgets: his animated octopus in *It Came from Beneath the Sea* (1955) had only six tentacles!) But one of the pleasures of fifties science fiction is that its low-budget technical effects seldom intrude and never dominate the development of narrative or cinematographic styles. In this respect at least, things have certainly changed since then. Today's science fiction, with the rarest of exceptions like *Alien* and *Robocop* (1987), is lavish but all too often facile; rich in profits for the film-makers, but meagre in returns for the audiences who, bereft of anything preferable, flock to them. Such films replace the search for meaning and reflection—which, as we shall see, characterized many fifties films—with an arcane kind of connoisseurship of effects. Some science fiction films could boast top box-office stars like Kirk Douglas, Christopher Lee and Vincent Price, whilst others trailed now justly forgotten bit-players. As the popularity of science fiction films grew by leaps and bounds, all the old acts were pressed into service—Dick Barton, Abbott and Costello, The Three Stooges and even the Bowery Boys met monsters, flew into outer space and encountered aliens.

It may be that every nation's movie style, its recognizable genre, is born out of trauma. British cinema seemed to be reinvigorated by the experience of the war and drew from that to produce its memorable studies of adversity and homelife under strain; French cinema developed its films of corruption and pessimism from the trauma of disgrace and the ignominy of defeat; Italy turned its recent history of poverty and the corrupt incompetence of central government to good effect in the neo-realism of the fifties. In America the atom bomb provided the major instance of such a seismic disturbance. It was of course in America that the bomb was designed, built and first tested; American airmen dropped this American bomb, under the orders of an American President. For Hollywood to tackle this theme was to touch an urgent and far-reaching set of concerns, and this science fiction did superbly well.

Science fiction films were born under the atom-bomb cloud and reflected that heritage as well as a multitude of other themes and fears: that individuals were being swamped by mass society, that nature was being forced to bear further and further incursions by science and technology, that man-made environments were eroding human sensibility, that machines were beginning to dominate human beings. The films were born in radiation fall-out and were then nurtured and transported by a space rocket. For the fifties was paradigmatically the decade which launched both the atomic and the space age. These technological feats offered different, and in some ways contradictory, impressions. Whereas the drive into space was exhilirating and offered a sense of potential, a flash of hope, and expanding horizons, the atom bomb became a symbol of images of destruction and narrow sectarianism. One made man loom large in the scheme of things; the other reduced him to scale, rendering him pitiful and meagre. The space race seemed to open new frontiers, and the imagery was of pioneers following in the tradition of Columbus. The atom race, on the other hand, seemed to suggest a voyage into the abyss. From such tensions and contradictions the 'Golden Age' of science fiction films was manufactured. It was to change, as well as to the social consequences of such change, that science fiction addressed itself.

In *The Day the Earth Stood Still*, the most immediate source of change is of course the arrival of aliens on planet Earth. Here aliens represent, indeed epitomize, 'difference': different values, different cultures, different histories. The question the film poses is simply, 'how do we deal with difference?' One answer, circa 1950, is by violence—not understanding. Force replaces reason. 'This creature must be tracked down like a wild animal: he must be destroyed', says one character about Klaatu, who has survived the twitchy soldier's bullet and escaped into the metropolis. Another of the city folk accuses Klaatu of being a communist from Moscow, while a third praises the Government's quick, aggressive response. 'What else could they do? After all, they're only people'; to which a voice quickly replies, 'No, they're not—*they're Democrats!*' Aliens, it is plain, are an irritant and provocation; like a virus, they summon forth our most potent defences. The alien must be held at a distance, then cornered, finally destroyed. It would be a simple task to read off from this movie all those familiar signposts of the fifties—red subversion, super-patriotism, Cold War obsessions, recoil from liberalism—and then consider the job of interpretation complete.

But there are other themes in the film which subvert the initial impression it gives, and which suggest how subtle a play can be fashioned from alien speculations. For one thing we—the viewers—know that the alien presence is neither violent (we heard it bring a message of peace), nor Russian (we witnessed it land). The knee-jerk response of the authorities is wrong. We know better.

In the second place, there emerges a subplot composed of a romance between Klaatu and Helen Benson, a female Earthling with the right qualities to draw our sympathy (she is an attractive woman, widowed by the war and left to raise a child alone). She is open, articulate, and warm whilst all about are losing their heads. To reinforce the validity of her approach, it falls to her to save the day by uttering the phrase (apparently repeated across high schools for months afterwards), *'Gort! Klaatu barada nikto!',* thus stopping the robot from destroying the globe after Klaatu is struck down a second time. The message? Apparently, even an emotional response can be a powerful deterrent. A softly spoken word can calm the savage beast. As Americans retreated from public issues into family life, and male and female roles were being strictly defined, are we to see here a reinforcement or a denial of the value of the private life? Love as the cure for hostility does not strike one as typical fifties apologetics.

For anyone receptive to the message, there is even a third dimension to *The Day the Earth Stood Still.* It is not one easily visible, since the scriptwriter kept it secret from cast and crew, intending it to be, in the fashion of the day, 'subliminal'. This theme places Klaatu on the side of the angels by equating him with Christ. Klaatu's adopted name as he goes 'among the people' is Mr Carpenter. He is crucified and resurrected, twice. Human bigotry against aliens, it is suggested, is the equivalent of the heathen denial of the Son of God. This allegorical flavour, though often shunned by Hollywood, did nevertheless appear in a number of science fiction films, amongst them *War of the Worlds,* where it is suggested that a Martian invasion would be mounted to teach us all a moral lesson.

There is evidently a connection between such films and the religious crusades which swept through the US during the fifties (remember Billy Graham?), but it is surely a deeply unsettling one. The connection is manifest once we contrast the elements of *The Day the Earth Stood Still* with the conventional religious epics that re-emerged during the same period, like *The Ten Commandments* (1956), *Quo Vadis*

(1951), *Ben Hur* (1959), and 'the miracle story of all time', the now-forgotten biblical bestseller *The Robe* (1953). We shall be exploring religious themes present in science fiction in Chapter 7, when we will be in a position to see how religious, and indeed military motifs, served to contextualize, orientate and control scientific research.

Turning now to the figure of the scientist, it is clear that he occupies a central role in *The Day the Earth Stood Still*, as he does in a great many other sicence fiction films of the fifties. He is introduced at a key moment in the movie, when Klaatu, having adopted the earthly name of Mr Carpenter, befriends Helen Benson and her young son Bobby. In a scene replete with political and historic imagery, Carpenter and Bobby visit the Lincoln Memorial in Washington. The diplomat from outer space is searching for a wise figure with whom he can communicate his message of warning and advice. He is looking for the mid twentieth century's Lincolnite embodiment of those American values of freedom, equality and good sense. The President of the United States? Surprisingly, no:

> *Mr Carpenter*: Bobby, who is the greatest man in America today?
> *Bobby*: Well, I don't know . . . the spaceman, I guess.
> *Mr Carpenter*: No, I was speaking of Earthmen. I mean the greatest philosopher, the greatest thinker.
> *Bobby*: Oh, you mean the *smartest* man in the world?
> *Mr Carpenter*: Yes, that will do nicely.
> *Bobby*: Professor Barnhardt, I guess. He's the greatest scientist in the world . . .

So, a scientist is the 'the cleverest man in town', and he doesn't reside in the White House. Carpenter seeks him out, knowing that he must gather together the brightest minds and the leaders of all nations if he is to deliver his message effectively. He and Bobby look for Professor Barnhardt not in his research institute or laboratory, but at his home. Together, the duo reach a book-lined study and a blackboard scribbled over with strange equations. This is clearly an intellectual's den. The door is locked, but the spaceman breaks in and leaves a message. A while later, Mr Carpenter is summoned to the Professor's house by an army officer, and the alien is finally able to communicate with a man of reason and calm good sense.

Though perhaps not obvious at 40 years' remove, there are a number of striking features of the encounter between Mr Carpenter and Professor Barnhardt. First of all, it is clear when Mr Carpenter meets his man that we are seeing that model of scientific integrity, Professor Albert Einstein.

Einstein in the fifties was certainly not an uncontroversial, everyday genius. Like Barnhardt in the film, he was a major intellectual figure who in his later years became scientifically and politically marginal. An alien? Well, Einstein himself was attacked in the House of Representatives in 1945 as a 'foreign agitator', hauled before the House Un-American Activities Committee, and subsequently described as a subversive menace by McCarthy himself. Barnhardt (played by Sam Jaffe, who looked perfect for the part) effectively breaks rank with the officials of government,

Mr Carpenter (Michael Rennie) meets Professor Barnhardt (Sam Jaffe), the Einstein look-alike and 'smartest man in the world', in *The Day the Earth Stood Still*.

who are trying to track down Carpenter, and with the military, who are stalking him, by eagerly accepting the spaceman's offer of a dialogue. Indeed, he is fired by curiosity, not rage or fear: 'there are several thousand questions I'd like to ask you', he says enthusiastically.

The Einstein–Barnhardt figure is controversial in other ways which are best described by contrast to other Hollywood scientists of the day. In representing the scientist in various roles and disguises, science fiction films of the fifties abandoned old stereotypes of the 'evil' or 'mad' scientist which had previously been a staple of horror and fantasy films. *The Cabinet of Dr. Caligari* (1919) featured a mad doctor with strange powers; *The Man who Lived Again* (1936) starred Boris Karloff as a crazy surgeon who finds a way to transplant a brain. Karloff reappeared in the same year as a mad scientist in *The Invisible Ray* (1936), and made the role his trademark in *The Man they Could Not Hang* (1939).

The mad or 'crazy' scientist only made the occasional appearance during the fifties, and in films that were either expressly absurd, like *Zombies of the Stratosphere* (1952), or played for belly laughs, like *Monkey Business* (1952), which featured the double-domed Dr Fulton. The Nazi scientist, an icon of concentrated evil, surfaced in many films of the forties; Bela Lugosi became strongly identified with the part through such films as *Black Dragons* (1942) and *Ghosts on the Loose* (1943). But there are no Nazi scientists in fifties cinema, though they are resurrected in the following decade in films like *The Yesterday Machine* (1963)—'how dare you report Adolf Hitler as a madman! He vuz a great cheenius!'—and *The Frozen Dead* (1967). Indeed, *Destination Moon*, which was loosely based on an early Robert A Heinlein novel *Rocketship Galileo* (1947), dispensed with the book's subplot of the discovery of Nazis camped on the Moon. A theme that had evidently lent topicality to the story in the aftermath of the war had already become redundant.

The paucity of mad or bad scientists in films of the period was matched by the rarity of perfectly heroic scientists, of men willing, as in *Target Earth* (1954), to sacrifice themselves to save the planet, or of figures using every resource to stem an alien invasion, as in *The Flying Disc Man from Mars* (1951). Like their darker counterparts, these figures had only a short screen life during the fifties and fell from view by the middle of the decade, by which time the image of the scientist as menace or saviour had lost all resonance. It was as inoffensive as it was irrelevant. this is not to suggest that scientists had lost their power on screen; rather (as we shall see in the following chapter) that such power was negotiated in movies and seen to be sanctioned by external or internal control. Indeed, one of the pleasures science fiction films offer is to speculate about the nature and extent of links between expertise, in its various guises, and democracy, between 'science' and 'the people'.

A world without science on screen was as inconceivable as a world without the new technologies was beyond it. Scientists were a functional necessity, and a familiar part of the landscape. They held power and knowledge in their grasp. They were not to blame for the mutation monsters which invaded fifties screens; in fact, no-one was held directly accountable, in some measure because the science fiction film refused on the whole to trade simply in the rhetoric of guilt and morality.

On occasion, to be sure, the experimental researches of a scientist went wildly astray, reminding us of the start of the monster pictures of old. Recall Dr Praetorious' words in *The Bride of Frankenstein* (1935):

> 'After twenty years of secret scientific research and countless failures,
> I have also created life . . . as we say—in God's own image . . . Science,
> like love, has her little surprises'.

Science may well have its surprises, but by the fifties the profession as a whole was not held culpable for tragedies that occurred in its name. As the eponymous scientist remarks in *The Quatermass Xperiment*, as he contemplates the death of two pioneering astronauts on a mission, 'every experiment is always a gamble; the unknown is always a risk'. At the end of *The Fly* (1958), the scientist who has devoted his life to building and testing a 'disintegrator–integrator machine' to teletransport living matter is killed by his invention. In an epilogue, which is seen again at the start of *The Return of the Fly* (1959), his son asks why his father is no longer alive.

Soothing voice off-picture: He died because of his work. He was like an explorer in a wild country where no-one had ever been before, searching for the truth. He almost found a great truth. For one instant he was careless . . .

Boy: That killed him?

Voice (with grave intonation): The search for truth is the most important work in the whole world and the most dangerous.

Boy: I like that. I'd like to be an explorer like him!

As we shall shortly see, one of the dangers the scientist in this film faced was working alone, and another was to commit himself so single-mindedly to the search for truth. But here, as in other films, it is made plain that research scientists—indeed, everyone—have to come to terms with the new power of science—and have to learn to harness it, live with it and above all, control it.

But scientists were also depicted as being in need of self-control. The most dangerous Hollywood scientist is the isolated boffin. Indeed, the trademark of the evil/mad scientist of earlier and subsequent years was his predisposition to working alone, buried deep underground, like a mole. By contrast, scientists in the fifties are literally as well as figuratively never far from the social world—from civilization. Their place of work is, of course, the laboratory, which is often set in a research institute. The laboratory is a social world of its own, linked by many channels to the world beyond; a complex, integrated space in which occur some of the key events in science fiction films, including the depiction of scientists at work.

What is a laboratory? The question is by no means simple. The laboratory is a building that houses scientific equipment and in which experiments are undertaken. It is a space, more generally speaking, in which certain kinds of activity (typically, research) are legitimate and other kinds are not. In this it resembles the space of the church (site of prayer) or the school (space for learning).

Laboratories are a comparatively recent feature of science. As historians of science have shown, the status of the laboratory as a public domain, in which legitimate knowledge could be witnessed and produced, evolved slowly during the seventeenth century in contrast and in opposition to the alchemist's den, which was a private place for the initiates or secretists who laboured at the furnace (Shapin and Schaffer 1985). A key feature of the laboratory is that the activities which take place within its walls are public rather than private. It upholds an ideology of public benefit rather than private gain, promoted in the language of humanity and consensus rather than of individuals and conflict, and of work undertaken for the good of humanity rather than for personal gain. In this sense, the laboratory connotes a physical site and also has a more abstract sense of being, a 'poetics of space', as the philosopher Gaston Bachelard once referred to it in his 1967 work of that title. This is even so—indeed, especially so—for research on radioactivity, the newer alchemy, as Ernest Rutherford dubbed it. The atom age was represented on screen by the mushroom cloud, mutant monsters, the atom itself as a miniature solar system—imagery which, according to Spencer Weart's *Nuclear Fear: A*

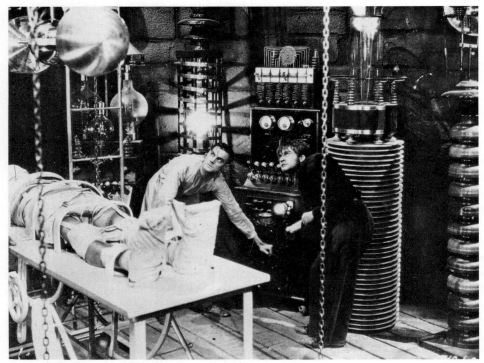

With the release in 1931 of *Frankenstein*, the science fiction horror film became instantly established as a crowd pleaser. This scene finds Henry Frankenstein (Colin Clive) and his assistant preparing to activate their monstrous creation. The laboratory is the site for illicit work and is kitted out in mad-scientist style: solid stone walls, dim lighting, bulbous rheostats and chains.

History of Images (1988), had deep, early roots. It was also portrayed and symbolized by the modern research laboratory.

In fifties science fiction films, the laboratory serves a role as central as the saloon bar in the Western, and the smoky office in the detective thriller. The horror film also portrayed laboratories, but these connoted different ideas—a place for retreat from the world, rather than a place to serve it; the activity of night rather than of day. In science fiction films the laboratory is fully integrated and cosy. It is the result of careful environmental and psychological (that is, ergonomic) planning: functionally efficient and comfortable, like a bungalow in the city suburbs. It is not a space for bourgeois comforts, like Captain Nemo's salon in the 'Nautilus', still less a place of sensual delights like Barbarella's fur-lined cabin. There is no upholstery, nor wood veneer. It promises neither rest nor retirement, but work. It has familiar-looking switches, dials and flashing lights—symbols of gleaming modernity. The fifties Hollywood laboratory also has some notable absences: we shall not see, and may well miss, the gleaming phallic columns, bulbous rheostats, and infernal electrical zaps. Nor shall we see that most central of mad-doctor devices, the destruct switch, which was often a huge lever, like a railway signalman's pole, which set off the inevitable chain reaction. The laboratory comes equipped with scientific appliances, but they look for all the world like domestic appliances, with their aluminium and formica; indeed, in many films, especially low-budget films, they *were* domestic appliances, like Dr Vornoff's sinister ray machine in *The Bride of the Monster* (1956) which is manifestly a photographic enlarger. Even in a feature film like *This Island Earth*, the technological marvel that Dr Cal Meachem and his colleague builds is assembled with the help of a manual from a DIY kit: '2486 parts for an Interociter'.

Walls are noticeably clean: whitewashed or tiled, and never, as in Dr Frankenstein's grotto, chiselled from rock or built of stone. This is important not only in ensuring that the laboratory *seems* domestic and familiar, that the activities undertaken within its walls have meaning beyond them, but also in ensuring that on those rare occasions when the scientist is left alone in his modern laboratory he is really surrounded by artefacts and symbols of society and culture. Environments

Dr Cal Meacham (Rex Reason) and his scientific buddy build an interociter using fifties DIY technology. They dress casually and work in clean, open surroundings. (From *This Island Earth*.)

which have no finery and nothing superfluous are alien environments, not scientific environments. Indeed, after Cal Meachem has built his Interociter and entered into communication with an alien intelligence, he is commanded by it to fly in an airplane to a research laboratory set up in the desert. The plane is a Dakota which the aliens have commandeered and rendered suitable for their own travel, and which they have made, in their alien way, comfortable. It has been stripped bare inside, painted white throughout, insulated from all outside noise, and programmed to fly without a pilot, the controls moving as if by their own volition.

In the laboratory white, starched coats are sometimes worn, but often dispensed with: Dr Meachem dresses casually for lab work, wearing pants and pullover. The laboratory is a signal to the audience that a certain kind of activity is underway or about to be introduced. The scientist is now surrounded by instruments, but these are the recognizable accoutrements of contemporary American society: dials, levers and lights, which service scientists, rather than the reverse. The scientist uses them as part of a carefully structured process of analysis, with established goals and procedures. *Phantom from Space* (1953) provides a good example. An invisible alien has landed his flying saucer in California, near a US observatory, and proceeds to kill some innocent picnickers. The killings alert the police and the scientists. The former must somehow start up a search for the intruder, while the latter must provide intelligence, and some understanding of the monster's habits and needs. The alien has been sighted: it is like a 'huge man in a diving suit', says one observer; and, 'I know this sounds crazy', says another, 'but he didn't have a head'.

The police are baffled though, under the weight of strong corroborative testimony, they find themselves edging towards an improbable solution. Then they finally retrieve the mysterious suit:

> *Police Officer 1*: Well, Hason, what do you say now?
> *Police Officer Hason*: Beats me. The descriptions check alright. This could be some kind of flying suit, high altitude equipment.
> *Police Officer 1*: Yeah, that's what I've been figuring, but . . . how do you explain all that stuff about the missing head?
> *Police Officer Hason*: Eh?!

From the linguistic difficulties and the halting speech, it is apparent that the policemen are stuck in a conceptual swamp. They need help, and go in search for it to the local laboratory. Once there, they hand over their 'flying suit' to a research team, who proceed to subject it to various analytic procedures. The appliance of science.

Every scientific age has its icons, those artefacts which manage to capture the spirit of the times. During the seventeenth century, telescopes then microscopes probed into the very large and very small, discovering worlds which enraptured poets and essayists from Milton to Swift. The eighteenth century for its part seemed transfixed by the display of electrical contrivances: in these the power of nature seemed to be tamed, then transformed, for man's pleasure and power. The nineteenth century witnessed the spectacular growth of the life sciences, typified by the development of evolutionary theory. The icon now was the primate skull, held at arms length by the scientist—T H Huxley perhaps, or his foe Richard Owen—who contemplated it with wonder. If Darwin brought man and his ancestry close together, the battery of machines, of high technology, has come in the twentieth century to separate the world of nature from those whose task it is to understand it. Indeed, devices such as centrifuges, spectroscopes, x-ray tubes, and electron microsopes seem as often as not to *create* phenomena for scientists.

In the public imagination of mid century, and to some degree even today, there is one instrument whose simplicity and power have associated it more than any other with scientific research in the atomic world. This is the Geiger counter, familiar from school science lesssons, and still produced for visitors to nuclear power plants to show that radioactive emissions are a fact of nature. Everyone knows that the Geiger counter gives a direct reading of radioactivity, and its bleep-bleep or rat-tat-tat sounds reassuringly familiar, like a heartbeat perhaps, or the call of an artificial satellite. In *The Thing*, the bleep-bleep of a Geiger counter provides the background music as well as the scientific semaphore.

When the scientist draws a Geiger counter from his toolkit he is signalling the onset of detective work as surely as did Sherlock Holmes when he produced his magnifying glass. In *Phantom from Space*, our two police officers have taken their mysterious space suit to scientists who then proceed to analyse it. They do so, inevitably, with a Geiger counter. Having established that it is indeed radioactive, the suit is now held at arm's length with two metal rods, then dropped into a lead-lined box, after which questions are posed. 'I want to know what it is and what it does' states the scientist. How radioactive? What weight? Does the suit burn? Does it dissolve in acid?

While this is going on, the police officers remain in the laboratory, but their presence is resented, as are their vague interjections ('Is it possible that . . .?', 'Suppose that . . .'). There is, then, not only a form of questioning that is appropriate in the laboratory, but also a style, a language. Moreover, as the strained relations between the police officers and scientists suggest, the space of the laboratory can be contested. It is primarily a site for scientific research, but representatives from government, state and local officialdom can intrude. What, then, are the relations that bind scientists together in laboratories and research teams, and how do these relations affect the integration of science in fifties culture? Science fiction films from the fifties give a number of answers to these questions, and it is to these that we now turn.

> *Exeter*: I represent a group of scientists who work for but one purpose. To put an end to war. Naturally, such a goal can't be obtained without experts of superior ability. Men of vision. Men such as you, Doctor, gathered here, exchanging information daily. Putting aside all thoughts of personal success.

Exeter is an alien scientist of advanced intelligence in the movie *This Island*

Earth who ultimately sacrifices himself for two human scientists; in so doing, he epitomizes a central theme in many science fiction films: the need for teamwork, for a scientific club, for the ideology of cooperation, loyalty and togetherness. The emphasis, even in a short monologue such as this, is evident: a group of scientists, of common purpose, exchanging information, leaving aside all personal motivation.

During the fifties, American men became concerned, so magazines and radio broadcasts suggested, with the decline of masculine endeavour, self-employment and the absence of a frontier to test their ruggedness. Individuals were being crushed by mass society. Men sensed that they were living unlived lives. Books like David Riesman's *The Lonely Crowd* (1950) and William H Whyte's *The Organization Man* (1956) documented the isolation of the self-made man and the decline of the inner-directed and self-motivated achievers, once spurred on by the Protestant work ethic, and the rise of men who were consumers and bureaucratic workers; the texts drew attention to uniformity, social-role playing and dullness. No inner drive propelled them, only the need to get along, flow with the tide, and preserve a space for their private suburban leisures. Ideology was pronounced to be redundant and inhibiting, and even political debate was judged to be inappropriate to what were largely seen, at least from the centre, as technical problems of distribution and production. As Bell's study of the decade proclaimed, it witnessed *The End of Ideology*; his subtitle significantly was 'On the Exhaustion of Political Ideas in the Fifties'.

In all of this the fate of the individual was continually debated, for he seemed, as Lionel Trilling noted in his 1957 study *The Opposing Self*, to be a basic unit submerged in mass society, but in need of protection and affirmation. Some voices were raised in criticism of the enforced consensus and drift to political quietism. Herbert Marcuse, for example, found behind all this an increasing centralization of control and a dangerous cult of technical rationality which liquidated the legitimate role of the individual and erased class conflict. The much-proclaimed age of tolerance was, Marcuse noted, underpinned and sustained by massive racial and sexual tensions as well as by economic inequalities; the conditions of tolerance were 'loaded', as he put it. But even Marcuse and his fellow critics were able to grant the scientific enterprise a privileged status. There conformity and consensus was thought to be harmless, even properly justified, since freedom was judged to depend largely on technical progress (Marcuse 1969).

The scientist in the 'Office of Scientific Investigations' makes this clear at the end of *The Magnetic Monster* (1953), in a thought redolent of the McCarthyite emphasis on communal values, when he points out that 'in nuclear research, there's no place for lone wolves'. The scientist is part of an extended nuclear family, working in the laboratory, a space which epitomizes collaborative research. In this, science fiction recalls many of the utopian visions of community, from Plato's Republic, through More's Utopia to Voltaire's Eldorado and beyond. In such communities, reason was the rule and the universal leveller. It acted as an intellectual glue, preventing controversy and easing tension. Since no conflict was intellectually plausible, none was actually possible. There are no conflicts at the tribunal of scientific reason, for

Nuclear research is dangerous, demanding cooperation and accountability. 'There is no place for lone wolves'. (From *The Magnetic Monster*.)

science (like reason) is an ideology of consensus. Science, epitomized by the laboratory, is a house with an open door, through which people (scientist or not) pass with their technical problems.

The scientist who rejects this ideology pays a severe price. Working alone, even when this is imposed upon the scientist, hampers and disarms him. In *The Day of the Triffids* (1963), a scientist is marooned in a lighthouse, from which he tries to discover the scientific basis of the triffids' existence. At first, it seems easy—'just like finding a weedkiller', he boasts. But isolated and lacking resources, he fails to make any impression. Indeed, the discovery that triffids dissolve in sea-water is made by accident, and it is left to a small girl to discover, again by chance, that triffids are attracted to sound.

The unhappy consequences of working alone is a message driven home in films which select the deviant scientist for particular censure. In *The 4-D Man* (1959), when a member of a research institute chooses to carry out illicit experiments, he is shown leaving the sanctioned space of the laboratory and establishing his research base elsewhere, in a secret hideout. His experiments backfire and though he develops a method of penetrating solid matter, neither he nor anyone else can control the process. He ages rapidly, proceeds to draw vital, recuperative forces from those around him, killing them in the process, and is finally shot dead by his fiancée. A scientist has placed himself beyond the community and suffered the consequences.

A less drastic but far more realistic expression of this appears in the English film *The Man in the White Suit* (1951), a rare example of satirical fifties science fiction, which serves as an acerbic reminder of what can happen when an innovation is bad for business. Alec Guinness plays a chemical genius who sets up a private experimental process in an enclave within a large industrial laboratory with the aim of developing everlasting cloth. This is an invention devised in private, and carried out amid an overgrown jungle of bubbling glass tubes and retorts in a corner of the communal laboratory, though the scientist's aims are clear-headed, humane and generous. The result is that the invention attracts the disapproval not only of the worsted interests, but also of the trade unions in the mill, and even of a washer-woman who poses the awkward question: 'What becomes of me when there's no washing to do?' This is not a film which despises technological advance, but one which makes clear that such advance must be negotiated and not planned in secret and then launched on an unprepared world.

The silicon jungle. Alec Guinness (right) has broken the cardinal rule of fifties science: work together for the good of humankind. As a result his lonely research, amongst glass tubes and retorts, brings misery to all in *The Man in the White Suit*.

The ideal scientific ethos is integration with the community, and presentation of even abstruse facts to the world: 'whatever scientific knoweldge this character has', says Dr Cal Meachem of an advanced alien scientist in *This Island Earth*, 'should be in our textbooks'. The penalty, in a harsh environment, when submission to consensual values or direct orders is paramount, is harsh. In *The Amazing Colossal Man* (1957), a brave officer disobeys orders by saving a fellow officer caught in the fall-out after an atomic test explosion. No matter his bravery—he is contaminated and meets an ugly death. Pride and hubris have led to a separation from the regulations and order of society, hence to the punishment of society itself, mediated by natural forces. The scientist who thought he could produce new forces is destroyed by them; the soldier who thought he could be a big hero grows into a monster and is cast out by wife, family and society.

It is not difficult to see that this pressure for (scientific and social) conformity weighed heavily on individuals in science fiction movies, as heavily, indeed, as it did for individuals outside them. In the majority of cases, films not only depicted a closed, ingrown and conformist world but seemed to laud it. Yet, there were exceptions, and *Invasion of the Body Snatchers* is one. It provides a subtle yet gripping study of the power and limits of conformity and consensus. Don Siegel's film, like others of his to follow—*Hell is for Heroes* (1962) and *Coogan's Bluff* (1968) in particular—is about the quality of life, more particularly about two social forces in conflict: the institutionalized, faceless bureaucracy and the isolated individual, first prompted, then bullied into rebellion. It is a film that draws elements both from the film noir, with its shadowy hallways and night-time sequences, and from the Western with its focus on law and disorder (see Gregory (1972)).

But *Invasion of the Body Snatchers* is a film which takes the theme and transposes it squarely into a fifties context. The film, for one thing, is given a science fiction twist as pods from outer space appear in a typical Californian town, Santa Mira (though one whose name calls attention to itself—mira in Spanish means 'look'). It is a middle-class everytown, where people relish spoon bread, tend their lawns, and meet for poolside barbeques; it is a place of gas stations, used car lots, advertisement hoardings, drive-ins, juke-boxes, and the sound of crickets at night. These pods then turn into replicas of human beings, replacing the originals in the process. The hero in this setting, in which all around him are being invaded and replaced, is a physician. In itself this does not guarantee opposition to gradual but relentless conformity, for the fifties witnessed the onset of a mounting tide of criticism of the medical profession (Burnham 1982). In fact Dr Miles Bennell is a somewhat reluctant hero, a rebel by default. He is only thrust into his role by the acquiescence of those around him, and is finally driven to revolt by the terrifying destruction of his fiancée.

This is a film both about the growth of an outer-directed pod society, but also about the consummate ease with which this society spreads. Not only is individuality swamped in a wave of conformity, but people are represented as acquiesing in this process. The film, though it has been seen as an attack both on McCarthyism and on communism, was planned by the director to be a more general exposure of

The pods—vegetable aliens which take on human appearance—arrive in Santa Mira, before being distributed throughout California. (*Invasion of the Body Snatchers* (1956).)

trends in fifties society towards instrumentalism and incorporation. As Siegel said, 'the majority of people in the world unfortunately are pods, existing without any intellectual aspirations and incapable of love' (quoted from Guy Brancourt's 'Interview with Don Siegel' (1970)).

The film makes this clear, as psychiatrist Dr Dan Kaufmann explains to Dr Bennell, 'it's a strange neurosis, evidently contagious. an epidemic, mass hysteria. In two weeks it's spread all through town'. 'What causes it?', asks Bennell. 'Worry about what's going on in the world, probably', replies Dr Kaufmann with assurance. This worry has brought about a kind of political paralysis, breeding a conformity which is like a living death. People become identical—like peas in a pod—regimented and sapped of any emotional individuality. This is a powerful film since it resonates with imagery and themes drawn from the witch hunts and tensions of the period. As Noel Carroll has noted, even the vegetarian metaphor 'literalizes Red-scare rhetoric of the 'growth' of Communism as well as the idea that revolutions are made by planting seeds' (quoted from Danny Peary's *Cult Movies* (1981)).

But the film also derives its power from the very haziness of its imagery, the

Dr Miles Bennell (Kevin McCarthy) trying to warn the world of the coming of the pods in the climax of *Invasion of the Body Snatchers.*

essence of which lies in the slow, alchemical transmutation of normalcy into strangeness and back into a different normalcy in an almost seamless process. The film refuses to fall into the traditional horror film formula: human beings are not possessed by or converted into monsters or beasts; nor is a bestial and primitive set of urges released in human beings. Rather, the effect of the pods is to take something precious *away* from people—the very essence of their personhood, their individuality of response, their emotional lives, their memories. 'There's no difference you can actually see', a character notes with puzzlement, 'but there's something missing. There's no emotion, *none*. Everything else is the same—but not the feeling'. As Susan Sontag said in 'The Imagination of Disaster' (1974), the soulless simulacrum 'has simply become far more efficient—the very model of technocratic man, purged of emotions, volitionless, tranquil, obedient to all orders . . . the danger is understood as residing in man's ability to be turned into a machine'.

As we have seen, this is a change which is easily made; moreover, it is presented as one which is welcomed by the newly formed pods. Reflecting on the changes occurring all about him, Miles soliloquises on the film's basic theme:

In my practice I've seen how people have allowed their humanity to drain
away. Only it happened slowly, instead of all at once. They didn't seem to
mind . . . All of us, a little bit, we harden our hearts, grow callous. Only
when we have to fight to stay human do we realize how precious it is to
us, how dear.

'Love, desire, ambition, faith—without them, life is so simple', one pod-man says,
as if welcoming the relief from the pain brought by this emotional baggage. This
signals the ambivalence of the response we, the audience, are perhaps intended to
share. The ideology of consensus has come full circle. Klaatu who arrived on Earth
with a message of integration and harmony established a dialogue with scientists
and left the planet with a resounding call for a 'United Nations on the planetary
level'. Siegel's film seems to find control and subjugation in the cooperation and
consensus which so typified American fifties society, and which served as an
ideology for science.

In the final frames of *Invasion of the Body Snatchers*, Siegel intended to throw
the dilemma to the audience itself. Miles Bennell runs out onto the highway, in
pitch darkness, and tries desperately to stop a passing car. 'Look you fools, you're
in danger. Can't you see; they're after all of us—our wives, our children, everyone',
he blares out his alarm for a heedless democracy. No response: the cars drift by as if
in a trance.

That is a chilling reminder of how the weight of consensus—deemed necessary
for a fast growing scientific establishment—could stifle and threaten the very roots
of democracy. In the next chapter, we turn to Hollywood's creation of a definite
image of the research scientist as a figure of power and authority but one whose activities need close supervision. The issue science fiction films of the fifties addressed
is which group is to provide that supervision: scientists themselves, government, the
military, or laypeople. As we shall discover, the question is by no means easy to
resolve, and turns, finally, on the character of the research scientist, on his (or her)
integrity, honesty and, of course, patriotism.

5

THE SCIENTIST AS REGULAR GUY

Fifties films present scientists at work in laboratory settings, upholding the ethics of cooperation, consensus and communalism. The laboratory is a privileged space for scientific research, and a key site in science fiction films; it is where scientist speaks science unto fellow scientist. But the laboratory maintains open doors; it is also a social space in which people come and go, and to which non-scientist outsiders are able to bring tasks and problems. These are first translated into scientific parameters, then carefully analysed, and finally solved.

What kind of people work in laboratories? Scientists, of course—but what kind of people are scientists? That is the question we shall be addressing in this chapter, and we shall see that, so far as science fiction films are concerned, the answer lies not only in the work scientists do and in the language they speak, but also in their human personalities and physiognomies. The fifties scientist looks, speaks and acts a part—and a very surprising one it is too.

The post-war period witnessed a remarkable and quite unprecedented growth in the size and scope of the scientific establishment, and hence of its public profile. The promotion and publicity surrounding scientific activity was to some extent reassuring to laypeople; after all, scientists can help to solve pressing social and technical problems, and their work lay at the root of the enormous dispensation of devices and gadgetry which equipped the American convenience home. But the reach of science and technology into the public domain had a darker side as well. Scientists, Oppenheimer had said, lost their innocence after the bomb was dropped on Hiroshima, and to many people the march of science brought doubt and fear. Are scientists building bigger and more destructive missiles? people wondered aloud. Are robots being invented to take over traditional tasks? What new and hideous life forms are scientific Merlins creating in their laboratories?

In the chapter that follows this one we shall be asking how Hollywood science fiction films treated questions concerning the control of scientific activity, and how they sought to integrate scientific ideology into patterns of traditional values. But first of all we shall explore in some more detail what kinds of images a cinemagoer of the fifties would have formed of the character of the scientist.

We have already seen that scientists were seldom depicted as working alone; or, if they were, they were invariably working towards a nefarious goal. Teamwork was the essence of the scientific compact. But what else can we discover about the work of scientists, besides its cooperative nature?

Science today is popularly associated with activities in the much-publicized fields of cosmology and particle physics; genetics and artificial intelligence research are also well reported. For the most part, the impression given is that the scientific cutting edge is applied to the very large or the very small. These were certainly themes of science fiction in the post-war period, but in moulding new images of the scientist fit for the Eisenhower era, Hollywood was committed to a somewhat different view of what was interesting and conventional in the activities of scientists.

Perhaps suprisingly, pure and esoteric research was often ridiculed as irrelevant or old fashioned. Indeed, the world of ideas itself was distrusted, and intellectualism became a negative rather than a positive virtue. This was explicitly so in anti-communist films of the decade, in which the sallow, bespectacled intellectual is pursuing a dangerous, un-American vocation to undermine the constitution, for which the only antidote is the home-spun virtues of the Bible, competitive sports, and the ideology of conformity and loyalty. More generally, Hollywood—in whatever genre it worked—depicted pure reason as not only aimless but also potentially dangerous. It was almost as if ideas themselves were liable to spread infection like science fiction spores.

Against this ingrained distrust, science in the movies had to bend to the practical, technological ethic—and be seen to be doing so. In *The Thing*, a splendidly raw and ferocious treatment of the theme of the monster brought back from its prehistoric slumbers, scientists divide into two opposing camps. One wants to defrost the alien creature that lies buried in ice, while the other warns that it may contain dangerous organisms and should be kept frozen. The second team is composed of non-interventionist and decidedly cerebral scientists; they feel it is better to leave the beast alone until 'more is known about it'. Some might think this a wise precaution, but the option is ridiculed by other scientists and by members of the US Air Force present on site. The cautious scientists are attacked for their passivity as well as their disinterest. Knowledge as such, knowledge for its own sake, seems to hold only a minority appeal, and to be part of a distinctly malodorous ethic:

> *Dr Carrington* (with great enthusiasm): Knowledge is more important than
> life. We split the atom!
> *Bystander* (sourly): That sure made everybody happy.

To pursue truth is to go in hunt of a chimera, to become possessed of a kind of demon spirit. Scientists, naturally, have as much freedom to think that as any other brainy figures from Hollywood central casting, but they also have to think to a purpose and within constraints. In *The Beast from 20,000 Fathoms* (1953), a scientist who has witnessed the havoc created by a prehistoric monster has trouble convincing a psychiatrist of the validity of his own testimony. 'Listen Doc, I'm not inclined to let my imagination run away with me', he says emphatically, and adds, as if to settle the matter, 'I'm a scientist'.

Scientists are generally the 'brains' in science fiction, but, as what might be termed 'brain movies' of the fifties showed, not all brains are equal. There are smart brains and dull brains. And there is, of course, Einstein's brain.

No sooner had the owner of this famous and valuable brain died in 1955 than his hemispheres were removed by Dr Thomas Harvey, a physiologist at Princeton, who proceeded to run a battery of tests upon them. Slicing the brain into sections, mounting them on slides, and peering at them through a microscope, Dr Harvey and his colleagues were searching for what made this organ such a prize specimen. Over thirty years later, Dr Harvey is still working on Einstein's brain, or at least those layers of it which have been preserved in glass jars like slices from a German sausage. Harvey recently admitted with a note of disappointment that his tests had failed to uncover the source of Einstein's genius in the configuration, size or physiology of his brain. 'To all extents and purposes', he said, 'it is a very ordinary brain'. As Roland Barthes noted in *Mythologies*, a collection of essays first published in 1957, Einstein's brain was a route through which the scientist was most frequently encountered in popular culture, and as a commodity it was used to sell products as diverse as fountain pens and insurance.

Compared with the indignities suffered by other cerebra in the fifties, Einstein's brain was accorded some notable degree of reverence. At least it was preserved. In science fiction movies, when brains were not being destroyed with chemicals or electricity or by the introduction of a flesh-eating bug, they were being split, sliced, transplanted, removed and unceremoniously consumed. Better still, and with more appropriate visual impact, Hollywood liked removing brains and then transposing them to glass dishes, whereupon they would float in a transparent fluid, attached to wires and tubes, and occasionally linked to a pair of floating eyeballs.

But behind every Gothic tale there lies a moral, and these dark stories of excavation and transplantation served as a caution against pure, concentrated intelligence. It is true that the fifties did not inaugurate brain movies; some of the earliest fantasy pictures played on this theme, with *Go and Get It* (1920) supposing that a dead convict's brain had been transplanted into a gorilla; and *The Lion's Breath* (1916) offering the more imaginative idea that the ferocious and courageous mind of a lion was transferred along with its brain into the frame of a meek, retiring civil servant. But such films were invariably comedies, and comedies without real punch; in the fifties, by contrast, the power of science was such that many would have thought brain to brain transplants not beyond the reach of research. Removing, studying and replacing brains was an idea whose time had come. But it was an idea thought quite deplorable in science fiction films, one which could only have tragic and violent repercussions.

In such films as *The Man without a Body* (1957), *The Brain That Wouldn't Die* (1959), *The Colossus of New York* (1958), *The Creature With the Atom Brain* (1955), *The Head* (1959), *The Manster* (1959) and *Fiend Without a Face* (1957), the brain never functions properly when removed from its natural host. It rebels or retires. At a time when the brain was viewed as a concentrated form of intellectual muscle, such films press home the message that a brain without its nurturing body, intelligence without a seat, is erratic, deviant and dangerous.

Yet if abstract and abstracted reason was held aloft for criticism, the power and reach of scientific explanation was not. On the contrary, it was emphasized, and treated with a reverence that is likely to strike us today as faintly absurd. Scientists were cautioned not to allow their imaginations to run riot or to gravitate to matters of purely intellectual concern. But their brainpower, when suitably regimented and directed, commanded assent and demanded respect. Scientists could and did use their brains to provide information, knowledge, intelligence and, especially, to give answers to questions.

> *Question*: 'I don't understand, Doctor'
> *Answer*: 'Don't you see, General . . .'
> *Question*: 'You don't mean . . .?'
> *Answer*: 'Precisely!'
> *Question*: 'But, then . . .'

And so it goes on, following a ritual that appears in dozens of films and serves at least three functions. First of all, such a dialogue helps to introduce a complicated piece of scientese—the official verdict. Next, it assists in setting out the legitimate authority of the scientist; and lastly, it establishes his domain of competence.

Not all scientific explanations are equivalent, of course. Some are far-fetched—literally—as in this exchange from *Plan Nine from Outer Space* (1956) in which an advanced alien offers an account of the thinking behind the 'solaronite bomb':

> *Colonel*: You speak of solaronite, but just what is it?
> *Alien intelligence*: Take a can of gasoline. Say this can of gasoline is the Sun. Now you spread a thin line of it to a ball representing the Earth. Now, the gasoline represents the sunlight, the Sun particles. Here we saturate the ball with the gasoline—the sunlight—then we put a flame to the ball. The flame will speedily travel around the Earth, back along the line of gasoline to the can—or the Sun itself. It will explode this source, and spread to every place that gasoline—our sunlight—touches. Explode the sunlight here, gentlemen, you explode the universe. Explode the sunlight here and a chain reaction will occur direct to the Sun itself and to all the planets that sunlight touches . . .

In other cases, something approaching a more genuine interchange takes place, as in this sample from another low-budget, low-life movie of the period, *The Giant Claw* (1957), which again illustrates how lines of authority shift during a typical question-and-answer session:

> *General*: Well, it's your dime, boy—what is it you want to show me?
> *Scientist*: How to shoot that bird out of the sky!
> *General*: Some new type of weapon?
> *Scientist*: No, with regular guns, bullets and bombs—anything you want . . . Now, I don't care whether that bird came from outer space or Upper Saddle River, New Jersey! It's still made of flesh and blood of some sort—and vulnerable to bullets and bombs.

General: If you can get past that anti-matter energy screen!
Scientist: Right! That's exactly what I think I've figured out how to do . . .
Now—this is a blow-up I had made of her bubble chamber photograph.
The chamber was bombarded by high-speed particles. The result? Notice
this hole—this gap right here. This gap is one of the most exciting and
significant recent discoveries in all science . . .
General: Atoms of matter or anti-matter!!
Scientist: Right. Now if this thing of mine works and we can get
close—real close—and bombard that bird's anti-matter energy shield
with a stream of mesic atoms, I think we can destroy that shield. The bird
would be defenceless then except for beak, claws, and wings. You could
hit it with everything but the kitchen sink.
General: We've got the kitchen sinks to spare, son! Do you think you
could do it?
Scientist: I'm not sure—but it's certainly worth a try!

It is in dialogues such as this that scientific explanation, and by extension the
validity of science itself, is put to work, judged for its efficacy, then accepted or
rejected. Scientists encounter a barrage of suspicion and disbelief—'Atoms of
matter or anti-matter!!'—and work assiduously to overcome it. In the clip above,
General and Scientist speak the same language, with a minimum of fussy and
technical content. At other times, the barrier raised by the idiom of scientific
language defeats laypeople. Scientists then follow Cal Meachem's tactic. In *This
Island Earth*, Dr Meachem is asked by a posse of journalists to explain his work. He
tries valiantly, before restoring to the familiar ploy:

> *Dr Cal Meachem*: I'm concentrating on the reconversion of certain
> common elements into nuclear energy sources.
> *Reporter*: Eh? What's that again?
> *Dr Cal Meachem* (smiling): Ha! What counts is *how I make it work*!

Making it work is one thing, but making it convincing is another. Scientists go to
great troubles to explain their work and its aims; on occasion, it is the validity of
scientific belief itself that they are required to articulate and defend, offering
audiences the opportunity to sense the power and limitations of science. Note the
astronomer's circumspection in this exchange from *Invaders from Mars* (1953):

> *Astronomer*: We do know this: there isn't enough oxygen on Mars and its
> surface is too cold to support life as we know it. There is a theory
> consequently that their cities are underground near some central core of
> heat or that they live in spaceships.
> *Woman doctor*: You don't believe that!
> *Astronomer*: I'm a scientist.

Which means what exactly? That he believes nothing or everything? That he trusts
anyone or no-one? As the dialogue proceeds, it is a young boy who guesses the

Science looks a convincing activity in this laboratory scene from *Invaders From Mars*.

correct mission of the aliens on Earth (recalling the insightful innocence of Bobby Benson in *The Day the Earth Stood Still*):

> *Woman doctor*: It can't be proved.
> *Astronomer*: It can't be disproved either. Could you disprove, for example, that the Martians have bred a race of synthetic humans to save themselves from extinction?
> *Young bystander*: Synthetic humans?
> *Astronomer*: My theory calls them mutants.
> *Woman doctor*: Mutants!? What would they want here?
> *Boy*: To conquer us!

Thus far in this chapter we have seen how the scientific community was called upon to police itself, and how scientists were required to offer explanations of their work and to orientate speculation to practical objectives. What of the character of the scientist? In an interesting analysis of the scientist in science fiction stories published from 1926 to 1950, Walter Hirsch has found that his image became increasingly tarnished during the period, until a reverse process began at the start of the fifties. The prestige of science then began to grow by leaps and bounds, and the figure of the scientist was polished up and promoted as never before (Hirsch 1958, Carter 1950). It is no great surprise to find that scientists themselves warmed to the new images being produced—images that glamorized them and reaffirmed the humanistic value of their creed (Barron 1957).

If science fiction literature reaffirmed the status and stance of science, the same is true of films during the fifties. Here we find that the scientist assumes greater and greater prominence in addressing and settling social issues, these having been translated into scientific problems. Moreover, the scientist sheds his cranky, eccentric brain, with its dishevelled imagination, and trades it for the kind of brain that is perhaps better organized, but essentially no different from any other intelligent brain. The scientist loses his 'mad' or 'bad' head and gets instead a standard issue, with all the conventional attributes.

This suggests, and it is indeed the case, that many of the accepted screen stereotypes of the scientist, endemic in Hollywood productions from the earliest, jittery days of silent movies, were being rejected. In their place there appeared a scientist who was barely recognizable as such, until he (or she) entered a laboratory and so assumed the role and guise of the researcher. This is remarkable, especially bearing in mind the reliance cinema has always placed on pictorialism to offer clues for the professions. The gambler wears a boot-lace tie, the preacher a black hat, the newspaperman a bowler, and so forth. Fifties science fiction offered no such sartorial clues for its scientists.

The scientist is, as a youth observes in one of the many teenager science fiction films of the period, 'a regular guy' (in the English vernacular, 'a good bloke' or 'a decent chap'). He dresses like Dr Average and even, in the idiom of the day, 'dates'. Scientists dance, drink and have fun. They are regular guys who have become scientists, not professionals who have become street-wise. They are human beings, and valued as such. They have human capacities and human frailties, and even tetchy and disagreeable scientists are prized. Dr Bernard Quatermass makes mistakes, is prickly and overbearing, but is bright and useful. He is a human being valued as a scientist. Science has not put nature under the control of humanity, but only of a sub-set of the species, composed of scientists and their allies. Science has not taken power from supernatural gods and placed it in the hands of mankind, but placed it under the trusteeship of scientists—suitably controlled and overseen, of course. The primary instrument of science is not the naked intellect, but the scientific intellect, trained and specified, in the exercise of which scientific values—not humanitarian values (though they can coincide)—are paramount.

If scientists are not immediately recognizable on screen, it is because they have become domesticated. More specifically, they have become naturalized. The scientist is no longer mittel-European, more middle-American; more likely to exude a pipe-smoking trustworthiness than an aura of cultivated knowingness. Better still, the Hollywood scientist smokes filter-tips, and since facial hair was about as popular as dandruff in the fifties, he is clean shaven. He is an Eisenhower boy like Dr Douglas Martin, 6 feet 3 inches tall, broad-chested and virile. In other words, a Brad, Chuck or John-next-door type, often played by a man like Richard Carlson, the most popular actor of fifties science fiction, who embodied 'the clean-cut intelligence that audiences demanded of their scientific heroes' (Baxter 1979). This was important since in other film genres, like horror and fantasy, the Anglo-Saxon name almost without exception denoted a sympathetic character, while a foreign name, especially when allied to a foreign accent in those xenophobic times, connoted a villain. (A more extensive discussion of this point may be found in Robert Plank's article 'Names and Roles of Characters in Science Fiction' (1961).)

In looking at the scientific man, we find in science fiction films that those aspects of the character that contribute to his scientific capabilities are cherished, at the expense of others, which can often atrophy and die. In this process, the scientist submits to various physical changes. He becomes smooth and polished. His processes of thought and expression are also domesticated. The scientist must learn how to speak plainly, and discard scientese and high faluting obfuscation. A lapse into abstract intellectualism in which language runs away with itself into narcissistic complexity is quickly taken up. Jargon itself is cause for censure. 'Why don't we all talk English', an FBI agent asks Dr Medford testily in *Them!*, before she has had the chance to finish a 'my theory is . . .' explanation.

The scientist as communicator. In this picture from the film *Them!* the scientist happens to be a woman and she needs to speak bluntly to the political and military authorities.

The art of communication was and remains the key factor in a world driven by ideological disputes, in a country fractured by political contest and rivalry, but in which consensus was at a premium. Personalized language was promoted by advertising and government propaganda and served to enhance community spirit. 'Your' local congressman, 'your' highway, 'your' newspaper—the idiom promoted the self-identification of the individual with the functions performed by others (often bureaucratic institutions). The hyphenated attributive construction was also deployed giving the appearance of fixity and fusion of different attributes. So the 'military–scientific' establishment was born, suggesting a unity, harmonizing contradictions, effacing differences, and bringing together the understanding of weaponry and its deployment (Marcuse 1972, Lowenthal 1961). (The description of Teller as the 'Father of the H-Bomb' served the same function, suggesting a family origin for technology, and conjoining the personal and impersonal.)

Science fiction films took language and communication seriously and explored it with great thoroughness, building up a picture where common language was a guarantor of consensus but always difficult to achieve (Plank 1959). Plain speaking is easier to recommend than to put into practice. The medium of science fiction itself, as we have seen, is poorly suited to introducing scientific explanations; they can seem intrusive or irrelevant. Moreover, a genre that aspires to being other-worldly, fantastical and awe-inspiring is sooner or later going to find language restrictive. This is why some of the best science fiction films such as *2001: A Space Odyssey* (1968) and *Silent Running* (1971) have proceeded with the barest minimum of dialogue. Films that have deployed dialogue in the conventional manner have often proved disappointing: the richness of the image is diluted by the banality of the verbal expressions accompanying it. 'We have lift off' hardly adds to the supreme spectacle of a rocket blasting into space.

The jargon of science fiction was often accompanied by an unemotional, flat and soulless voice, exemplified by the alien 'It' which we discussed earlier. But there were mellower tones to listen to as well—musical ones for example. Music served as a powerful and evocative counterpoint: eerie and atmospheric, it worked to highlight and yet smooth over the abbreviations of language and jargon. Language was normalizing and Earth-centred, while music often seemed distant. But in fact, scientific language itself was often held up for criticism, as being archaic and pretentious, and an inhuman imposition of a kind. (One wonders indeed whether the struggle against scientific staccato and the rigid formula was not linked in the fifties to the war against communism, for in both cases the threat was perceived to lie in the attributes of the slogan, with its rigid, declamatory, effacing style.)

Speak in plain language and there is no longer a cover for hypocrisy, deceit or propaganda in foreign-sounding words. Common sense was something the Hollywood regular guy possessed in great measure. If jargon erodes human values, common sense expresses them. It is a veritable weapon in the struggle against aliens and alienation: a link to community and togetherness. The victory over the monster in *Them!* is the direct consequence of scientists and military officers playing it not 'by the book' but by using their native wit and resources. To accept the scientific method as the only path to understanding the world is to risk losing the power to impose order on it; to retain that, common sense and traditional values are necessary. Indeed, lacking these, all parties can become trapped in ideological configurations, quite unable to break out into a common area of discourse, unless it be by brute force. In *Plan Nine from Outer Space*, communication breaks down totally when two military officers try to fathom out a highly intelligent, but alienoid, scientist:

> *Colonel Edwards*: Why is it so important that you want to contact the governments of our Earth?
> *Alien Scientist*: Because of death. Because all of you on Earth are idiots!
> *Jet Pilot*: Now you just hold on buster!
> *Alien Scientist*: No—you hold on! . . . Your scientists stumbled upon the atom bomb—split the atom! Then the hydrogen bomb, where you actually explode the air itself. Now you bring the total destruction of the entire universe, served by our sun. The only explosion left is the solaronite.
> *Colonel*: Why, there's no such thing!
> *Alien Scientist*: Perhaps to you, but we've known it for centuries. Your scientists will stumble upon it as they have all the others. But the juvenile minds which you possess will not comprehend its strength until it's too late!
> *Colonel*: You're way above our heads!
> *Alien Scientist*: The solaronite is a way to explode the actual particles of sunlight!
> *Colonel*: Why, that's impossible . . . a particle of sunlight can't even be seen or measured!
> *Alien Scientist*: Can you see or measure an atom? Yet you can explode one! A ray of sunlight is made up of many atoms . . .
> *Jet Pilot*: So what if we do develop this solaronite bomb—we'd be an even stronger nation than now!
> *Alien Scientist*: You see! You see! Your stupid minds! Stupid! Stupid!!!
> *Jet Pilot*: That's all I'm taking from you [he knocks the scientist to the ground].

The frustrations that boil over in this scene are caused partly by the language of the alien scientist, partly by the threatening tones, and not least by the barrier of incomprehension that separates the expert who understands natural phenomena (albeit incorrectly in this case) from those that think only of controlling them. In other cases, scientific language, while not provoking such violent reactions, is seen as cold and heartless. It is a blunt instrument, ill-equipped to prise meaning from subtle, human interactions, as the following exchange—from *Glen or Glenda* (1953)—suggests:

> *Psychiatric Expert*: Most of us have our idiosyncrasies . . .
> *Police Inspector*: This fellow's was quite pronounced.
> *Psychiatric Expert*: Yes, but I wonder if it rated the death warrant it received . . . Let's get our stories straight: you're referring to the suicide of a 'trans-ves-tite'?

Police Inspector: If that's the word you men of science use for a man who wears woman's clothing, yes.
Psychiatric Expert: Yes, in cold technical language, that's the word, as unfriendly and vicious as it may sound.

Once the scientist has assumed proper proportions and a human face, his work begins to carry recognizable intellectual gravity and social implications, as for example in armaments and warfare, as we have seen. As a result of this, and of the common idiom in which communication is now undertaken, it becomes possible—indeed, legitimate—for everyday characters to query the direction of research and anticipate the consequences of new technologies such as spacecraft, power sources, and robotics. Ordinary people on screen, and by extension off screen, can (or so it seems) participate in decision-making about the role and shape of scientific and technological developments.

In some cases, we are invited to take or reject the explanation offered by an expert figure, a decision we, as audience, can often make from a privileged vantage point, having available to us far more information than the expert himself. This is emphasized by the presence of a third party, standing in, as it were, for the intelligent non-partisan man in the street, analogous in some respects to the third party in some scientific dialogues, such as Galileo's. These exchanges are seldom meant to exclude participants, but for the most part serve to import outsiders, including audiences, into the texture of the film. The vernacular and the dialogue format both work to ease listeners into the scientific, or scientific sounding, frame of reference. Here is an extract from a discussion between a psychiatric expert, a patient and an interested party who treats the explanation like a special offer at the supermarket. It is taken from *The Bride and the Beast* (1958), a piece of absurd pseudoscientific hokum, it is true, which only goes to emphasize the ubiquity of scientific-sounding rationalizations even in bad films.

Psychiatrist: Back, back to the endless reaches of time. Back, before your entry into this life as you now know it. You will speak of that life as you see it. Go on to your journey. What is your name?
Laura: Laura Carson.
Psychiatrist: Is that all?
Laura: Married yesterday to Dan Carson.
Psychiatrist: Go on . . .
Laura: Angora sweater was such a beautiful thing . . . Soft, like kitten's fur. Felt so good on me. As if it belonged there. Felt so bad when it was gone.
Psychiatrist: Dan, do you realize we've just witnessed a portion of your wife's previous existence, and her death in that previous existence? . . . Her talk about maribou, Angora, and fur-like materials—there's definitely a connection between them and her dreams. I believe that it's derived from her past existence.
Dan: Aw, c'mon. You don't really believe she was a gorilla!

Psychiatrist: All the evidence points to it. Her fixation for fur-like materials comes from that fact.
Dan: Sorry, Doctor, I just don't buy any of this.
Psychiatrist: Well, you have a right to your own opinion.

The moral here, as in other science fiction films, is that we may be safe in the hands of the Hollywood scientist but we must always be on our guard. One of the reasons for this is that the scientist, prodded and bullied from various quarters, as we shall see in the next chapter, finds himself in an uncomfortable position. He is liable to crack under the strain, and even revert to earlier type and go crazy.

The fact that science is portrayed as alarming and reassuring, as of great benefit but potentially of great harm, that scientists are regular guys but can all too easily move into another gear—all of this provides a source of deeply textured confusion. Scientists themselves are confused by their newly appointed tasks; the military is confused by the requirements to protect (Earth and its people) and destroy (aliens); and people caught in the middle either occupy an unsettled and uncomfortable place, or see clearly from their vantage point where others, with their vested interests, are floundering in an ideological fog. Indeed, what horror (and audience pleasure) there is in science fiction movies lies in the traditional scenario of witnessing, belief, disbelief, and final recognition of threat.

Because scientific language and values are so often subsumed by other more cherished conventions, so it is that ordinary people—the bearers and retainers of those conventions—are often located in a curious space. The first thing one might notice about the status of people in science fiction films is that they are often undifferentiated into classes or castes; there is a levelling of class distinctions in favour of a language of 'the people' or 'citizens' or 'local folk'. There seem few evident distinctions drawn in the lives or enthusiasms of the worker and his boss, the automobile owner and the petrol pump attendant. Instead there is collusion and alliance. People, as we noted earlier, were apparently invited to enter into dialogue with science and technology, deploying a common and accessible language.

In some cases, laypeople seem to have a special grasp on reality. It is a convention of a number of science fiction plots—at least of alien invasion plots—that one or a small group of homely, simple folk are the first witnesses to the monster, or UFO, or unaccountable phenomenon. No-one believes them, and it is only at the end of the story that the disbelievers are proved tragically wrong. (The audience, of course, occupies the vantage point: it knows that the witnesses are correct.) There is then a strand of science fiction which integrates the public, adopting its viewpoint, its language and its values.

By way of total contrast (and there is only a very small grey area in between), members of the public can be painted in far darker and more menacing tones. This is not especially surprising, since the picture of fifties America fast achieving an alarming degree of conformity had overtones of totalitarianism and menace, captured in Riesman's master metaphor of the 'lonely crowd', and described as a threatening, volatile presence in Hannah Arendt's *The Origins of Totalitarianism* (1951), which saw the mass as a mob.

Consider again this piece of dialogue, from *Phantom from Space*, which follows a series of sightings of an alien. Three local townsfolk have reported to the police the appearance of a 'huge man', apparently quite headless, in a diving suit. Earlier, we cited it to give an example of non-scientists struggling with a mysterious piece of evidence. We can now see that there is another dimension to the exchange, which suggests that the public only has a frail grip on reality.

> *Police Officer 1*: Well, Hason, what do you say?
> *Police Officer Hason*: Beats me. The descriptions check alright. This could be some kind of flying suit, high altitude equipment . . .
> *Police Officer 1*: Yeah, that's what I've been figuring, but how do you explain that stuff about the missing head?
> *Police Officer Hason*: Eh! We can discount that. The people were frightened. They panicked. It was night. No-one really took a close look at them.

The vision of laypeople suggested in this dialogue is disturbing and repellent, and it is one which is found all too often in science fiction. It is as if the elevation of the scientist to a position of epistemological advantage entails the reduction of people to insignificance. They know nothing. And in many films they *are* nothing. Think of the way people are summarily dealt with by aliens: not just killed, but exterminated. In *War of the Worlds* the death ray reduces human beings to a pile of ash; *The Quatermass Xperiment* presents us with an astronaut returning to Earth who is infused by a mysterious spore and decomposes before our eyes like a rotting vegetable. In the Japanese film *The H-Man* (1958) people simply liquefy. Always, it is the alien which is endowed with power to disperse or exterminate human bodies: in *Teenagers from Outer Space* (1959) 'they blast the flesh off humans'; in *The H-Man* 'touch it and only your clothes are left'.

In *The Fly*, the equivalence of man and non-man, the reduction of human values to nothing, is stated and restated with bold and repugnant detail. A scientist with the brain of a fly, a fly with the head of a scientist: no difference. Total equivalence, and this process of total and literal dehumanization is enforced in science fiction films in general by the overall lack of depth of character given to aliens and monsters. For, with only a few exceptions, camera angles and direction offer the perspective of a neutral observer, insisting on the same view for alien and human, projecting a kind of visual equivalence between the normal and the freakish, from which follows, as Frank McConnell has noted, 'a devastating reduction of humanistic perception' (McConnell 1970).

Invisibility brought little horror to fifties cinema, and it was seldom conveyed on screen. Instead it is dismemberment which brings the ultimate horror, for it dissolves the most precious attribute of narrative fiction—the unity of character. The power and reach of science fiction fantasy lies in some measure in its ability to interrogate the character of character. The shrinking, growing, dismembering self is reminiscent of the grostesqueries of Hieronymous Bosch and Salvador Dali, of the scattered human objects in the universe of Jacques Prevert and Andre Breton. Indeed, magnification, diminution and reconstitution are prominent themes in

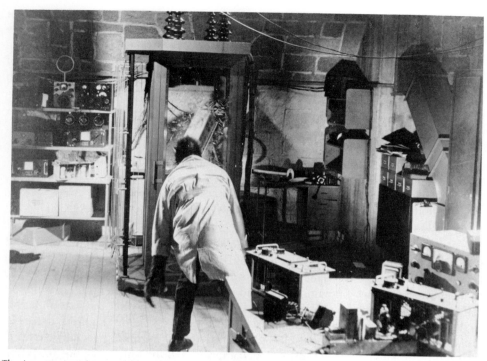

The insect man destroys the machine that created him, in *The Fly*, a film that offers a perplexing set of moral and scientific issues.

satirical and allegorical fantasies of old, where they serve, one might say, to relativize perception.

But science fiction films of the fifties were unusual in giving the change from whole person to pieces no immediate meaning, or rather of not slotting it into any kind of easy teleology. Change *is*. It leads to no particular destination. That is why it is terrifying. Gone is the notion of self as unitary, stable, undivisible, and continuous. There are now images of personhood where all that is solid melts into air. Bodies here reverse the process of production of Frankenstein's poor creature who was literally constructed from disintegrated selves, remade from corpses. Those caught in atom-bomb clouds, or mists, are destined to become corpses—inorganic. Thanatos has replaced Eros.

Other films present us with the undead, the unliving, the unperson: anonymous creations living in a twilight world, or which can be brought there by a horrifying embrace which mysteriously, but with consummate ease, transforms wretched human beings into automatons, eats their brains, obliterates their personhood. The people 'taken over' in *Invasion of the Body Snatchers* are perfectly happy with their new existence: they are more efficient, capable, and better.

On a larger scale, people, often a mass or crowd, subsist perpetually in an uncomfortable limbo, on the verge of wild and blind panic, ready to assume the proportions and intent of a Dickensian rabble. 'Has everyone in this fool town gone crazy?' someone cries in *The Blob* (1958). At the mention of impending danger or doom, the crowd becomes hysterical, then cowardly and greedy, and finally emerges as a venomous, vengeful mob. This is certainly the transformation undergone by the people left on Earth in *When Worlds Collide*: they are alarmed and then riotous when they learn that the rogue star Bellus is on course to destroy their planet.

The common people, whose idiom and perception were so reliable and steadfast in other movies, are here feeble. They are an easy target for mass hypnosis, as in *Invasion U.S.A.* (1952), and bow down before television or radio, or some other source of propaganda which has fallen into the hands of aliens, fifth columnists or asiatics. Faced with perplexity or fear, they fall into a silence broken only by grunts and groans. The only access they have to centres of command, and the only access these centres have with them, are via the newspapers and radio. The figure of the journalist is key, but since the crowds are so volatile and unreliable, the journalist is kept in the dark. Reporters can blow the lid off controversy and threaten official secrecy. 'Do you think all this hush-hush is necessary?', a newspaperman asks Dr Medford in *Them!*. 'I certainly do', she replies. 'I don't think there's a police

When Worlds Collide made life difficult for viewers. Would you kill your best friend/mother/child to get a ride to safety and escape certain death?

force in the world that could handle the panic of the people if they found out what the situation is.' More than once, a hectoring pressman is mistaken for the enemy agent or alien and almost blown apart by a howitzer. 'What they don't know won't start a panic', says scientist Dr Quatermass to a policeman, who immediately suggests putting out 'an anti-panic statement'.

The slightest murmur, the most mundane detail, is enough to fire anxieties and paranoia of the crowd, providing crowd scenes as memorable and as stylized as those in earlier films by Sergei Eisenstein and King Vidor. Not surprisingly, the result is not to settle the confusion generated by ideological conflict, but to magnify and loosen it. Driven by a feeling of powerlessness, or simply abandoned to their fate, the crowds in many science fiction films simply disappeared. The audience in these circumstances seldom had their sense of disorientation, powerlessness and confusion settled for them. Whilst other genres of movie had structure set deep within them, and came equipped with an ending which tied together any out-standing strands of the plot, science fiction films were often happy to leave outstanding issues of character development, fate and the future hanging without resolution or evident effect.

Characters are trapped in cross-currents. The only faintly—and evidently ironically—reassuring message was that science which had created the problem in the first place, say by nuclear testing, was perhaps capable of providing the solution to it in the future. The dialogue which ends *Them!* offers a world in which the future course of humankind lay still in doubt, and possibly beyond immediate control:

> *Army Officer*: Pat, if these monsters got started as a result of the first atomic bomb in 1945, what about all the others that have been exploded since then?
> *Pat Medford* (scientist daughter of Dr Medford): I don't know . . .
> *Dr Medford*: Nobody knows. When man entered the atomic age, he opened a door into a new world. What we'll eventually find in that new world, no one can predict . . .

At the end of *Invasion of the Body Snatchers* the camera picks up a horrified expression in very close-up to Dr Miles Bennell's face, establishing both a vicious implication of the audience in the horrors on screen, and a propinquity with his plight. He turns towards the camera, assaulting us with his outstretched finger and screams: 'You're next, you're next, you're next!' This, as Siegel recognized, was one of the most dramatic endings in the history of the cinema: 'the curtain came down and you were in a state of shock because you didn't know whether the person sitting next to you might be a pod' (Bogdanovich 1968). But as Siegel explained later, the film company felt the original ending was too distressing; the film, as released, opens with Miles in a hospital claiming to an incredulous psychiatrist that creatures from outer space have taken over the town of Santa Mira, California. Miles then proceeds to tell his story in flashback, and the film ends with him concluding his story to a now totally disbelieving psychiatrist, who is preparing to

The human group: a panic waiting to happen. Dismay, awe and terror are on the faces of those that set eyes upon the aliens in *It Came From Outer Space*.

have him taken away. At a critical moment, another physician enters to say that a man has been killed driving a truck filled with strange pods; the pyschiatrist now believes Miles and the film ends with him telephoning the FBI.

The Thing carries other implications, and refuses to be either reassuringly optimistic or comfortably sad. Mankind is saved, but only for the time being, claims a broadcaster haltingly. 'I bring you a warning . . . to every one of you listening to the sound of my voice . . . tell the world . . . tell this to everyone wherever they are . . . watch the skies . . . watch everywhere. Keep on looking! Watch the skies!!!' Here is the most powerful of endings. Shouts, like ellipses at the end of a foreboding sentence.

Endings are crucial in science fiction films, or indeed, any films that touch on serious issues and dilemmas. As Billy Wilder's *Fedora* (1978) puts it, 'Endings are very important. That's what people remember. The last exit. The final close-up'. The need for dramatic closure and resolution has often spurred the film maker to suggest answers, and tie up issues—comforting and comfortable. This, however, is to militate against the intellectual approach, which is to not dramatize standing issues but to advance and contribute to those issues and open them up for debate, as Ian Jarvie has pointed out in *The Philosophy of Film* (1987). The fantasy aspect of science fiction, and a contribution to its subversive potential, lies in its essentially open character and in its opening ability. It actively disturbs, denying the solidity of what previously passed for the real and the consensual, tearing reality apart. To end by sewing it back together again would be neat and tidy, but would counter science fiction's current of violation and contravention. Science fiction endings are best when they are sudden, dramatic and open

6

THE REAL WORLDS OF SCIENCE FICTION

Good science fiction works, but how does it work? Largely by retaining some contact with the real world. This is something which science fiction has managed to do surprisingly well, and part of its attractiveness derives from its ability to deal with issues of the day, snatching headlines from newspapers, radio, or lately television, and working them into grandiose fictions. By so doing, science fiction has managed to keep up to date and to reflect public fears and aspirations, as well as offering insights into topics such as science and technology, war and peace, social organization, and education. In fact, there seem to be no substantial issues that science fiction has left untouched.

Yet, if the genre can lay claim to topicality, it must by the same token accept the charge that it dates remarkably quickly. Newspapers have a very rapid transit from the presses to the wastepaper basket; a good deal of science fiction is barely less ephemeral. It probably has a half-life of about five years, after which period its potential audience has halved—unless it has succeeded in latching onto an issue of lasting interest or it has managed to receive quickly the accolade of the 'classic work'. The classic piece of science fiction lives on either by virtue of its inherent artistry or because of the richness of its vision, or as a consequence of both. These seem to be the criteria which can save science fiction from the unforgiving passage from pulp to print to pulp (or from celluloid to cutting-room floor).

These remarks suggest that science fiction has at least two features that can make it work, and lead to its survival. One is its bearing to the real world, the other is the skill with which it extends and extrapolates issues from that world. This second ingredient—artistry—is the more difficult to pin down but perhaps easier to recognize. So far in this book we have sought to trace the element of original imagination, or satire, or suspense, or skillful play with ideas that sets classic science fiction apart. Without ignoring lesser works, and paying perhaps more attention than is customary to samples from the dense world of 'good-bad' science fiction, we do acknowledge that without consummate artistry of one kind or another, the fantasy of science fiction lacks impact, interest and reach.

But without contact with reality, a fantasy has no evident meaning and will generally fail to engage readers and audience. In this chapter and the next we shall be dealing with the question of science fiction's relations with the real world of politics, ideas and power. Our discussion will first of all consider how the genre has approached issues in the real world, and then consider its radical, even subversive, features, particularly its lack of a moral context for man and alien in films of the fifties. Then, following on from the ideas set out in the earlier chapters on Hollywood's depiction of scientific work and of scientists, we shall examine in some detail how the difficult and evolving real life relations between the community of scientists and the military complex were represented on screen. The chapter which follows this one will consider the relations between science and the church, and focus particularly upon the treatment of religious themes in science fiction.

Taking the realism of science fiction first of all, we think it may be useful to try to avoid some of the rather limiting debates that have developed over the past 20 years around the category of 'realism' and what is termed 'the realist text' (a broad category which includes artefacts as diverse as a novel by Emile Zola, a newsreel documentary, a newspaper photograph, and a piece of advertising). These debates have addressed the question of how it is that certain works manage to be realist or realistic; how they manage to be successful, that is, in rendering the real world as if all they were doing was to open a window onto it. Various devices have been noted as effective in different media, ranging from the use of deep focus cinematography in films to the inclusion of the telling, but apparently irrelevant, detail in novels.

In our view, these debates have been enormously productive and have led to many fresh appreciations of the hidden artistry of pieces of work which seem (typically of realism) not to have been manufactured at all. Nevertheless, one drawback of these interpretative strategies has been that on the whole the essence of realism has been sought—and found—in the body of texts themselves. In our view, this concentration on the structure and content of individual works misses a key element, namely the audiences for them. It is self-evidently the case that the individual who reads a book, like the crowd that views a film, is not an inert consumer of messages and meaning, but an active being who plays a great part in deriving sense from reading or viewing. The audience, in other words, plays a major role in establishing the realism of texts. We should perhaps alter our frame of reference accordingly, even shift our terms of analysis, and talk not of realism but of verisimilitude, which suggests the important function of third parties in creating realistic effects out of what they see, hear and feel. Or better still, since this would seem to replace an inadequate term by a rather ugly one, we should prefer to think about the *believability* of films and books.

It is surely the case that film-makers and audiences, novelists and authors, have seldom, if ever, striven consciously to create 'realist' texts. On the other hand, they have sought to make their words and works believable. This is particularly so in the domain of science fiction. Occasionally, history itself has provided some help. For example, once Hollywood science fiction became concerned with native American themes it was able to have a greater direct impact and become far more believable. This it did once it had begun to nurture a fresh generation of writers independent of the European tradition established by authors of the rank of Mary Shelley, H G Wells and Jules Verne, and once it had recruited American actors to replace figures such as James Whale, Boris Karloff and Bela Lugosi. We saw in the last chapter that the scientist in fifties science fiction was naturalized and domesticated;

What is believable? If you have never seen a monster in real life how do you go about designing one for the movies? This is *The Creature from the Black Lagoon.* Maybe its credibility depends more on the reactions of the cast than on its physical appearance.

this would not have been possible without production teams in Hollywood which had sunk deep roots into American culture (or what passed for it in tinseltown!) So the historical development of the industry laid the basis for an important phase in science fiction cinema as well as for an output of films which totally overhauled the representation of science and technology in relation both to earlier depictions and to rapid changes in the real world.

The motifs of science fiction are without doubt often esoteric, fantastical and implausible, so to suggest that they are anchored to the real world needs a little clarification. We do not intend to imply that science fiction offers the same slant on the world as a news item on television or a newspaper column reporting a political speech. Tales of life in the future are projected forwards from the present to show varied, colourful and active beings. In a similar manner, tales of unknown worlds extrapolate from what is known and possible to what is unknown but not impossible, producing a fictional cartography bearing the same relationship to the real as Dante's world bore to that described by Aristotle. It is loosely and recognizably based on the original, but is more poetic, exotic and lively.

This subtle relationship of difference and continuity—which recalls the greatest satires of Jonathan Swift and Voltaire—holds throughout science fiction and applies whatever its peculiar subject matter. Aliens from the wilder reaches of the cosmos, for example, look, speak and behave as we might expect them to (with important exceptions, of course). Moreover, their creators on the page or screen for the most part encourage us to view them in Earthly rather than alien terms. Aliens are more (or less) intelligent, wise, and martial *than us*. Even in the world of fantasy, where the speed of light can be exceeded with impunity and black holes are minor disturbances along a spaceship's path—even in this make-believe, law-breaking world, man remains the measure of all things. As we shall see in the next chapter, herein lie some of the most traditional features of the genre and some of the links between space-spun science fiction and the Earth-bound traditional and religious values.

It would be easy to see science fiction as irretrievably trapped in anthropo-morphism, and hence as part of an age-old pattern of human domination of the Earth and the universe. In fact, although this man-centredness is reiterated in films and books, there are other voices to be heard. There exists a critical science fiction which has successfully undermined our comforting notion that the universe has been created around us solely as our playground, and it is to this counter-current that we now turn.

The idea that science fiction as a genre could articulate critical responses to contemporary ideology may perhaps seem preposterous. A historian said that the science fiction film 'derives from the comic strips and endorses the political and moral climate of its day' (Baxter 1979). In the fifties, however, critics detected other, far more powerful and serious responses. In 1954, for example, L W Michaelson surveyed recent science fiction and concluded that out of 60 stories more than 95 per cent included an explicit declaration against war (Sobchack 1987). This is not in itself surprising: politicians are in favour of peace, priests declaim against sin, and science fiction authors are against war. The conventionality of the stance is all the more evident when it appears that it was for the most part a blunt attack on unspecific targets ('war', 'casualties') rather than institutions (the government, the military).

Nevertheless, even this somewhat restrained radicalism did constitute a sharp break with earlier science fiction, whose popular heroes Buck Rogers and Flash Gordon strutted about the cosmos as explicit champions of the values of capitalist free enterprise, patriotism and aggression.

80 per cent of the stories that Michaelson analysed went a step further and located a specific target. Here war was depicted as the outcome of the direct manipulation of power-hungry rulers, or armament manufacturers, or international banking syndicates. 75 per cent of the stories declared against racial prejudice, either casually or directly. On the whole, the typical alien arriving on Earth could be expected to rail against all manner of human conventions and practices, amongst them meat eating, class distinctions, labour exploitation, unemployment, illness, soap operas on television, and even Hollywood B-movies. 'In fact', Michaelson

concluded, 'the overall implication of the majority of science fiction stories is that if we had one molecule of sense we'd grab the first available saucer ship to good old Planet X and never come back to this miserable, run-down, slum tenement called Earth again'.

In a later study, Michaelson considered that since 1951 there had been a tendency to prohibit pessimistic science fiction, a tendency which he went on to describe as inconsistent and more or less self-imposed by editors grown weary of the gloomy, critical, philosophical tales of science fiction. But this seems something of an exaggeration. After all, there were a number of popular stories that encoded radical, even subversive, themes. Cyril Judd's *Outpost Mars* (1952) described a struggle between idealists who had colonized Mars and drug manufacturers who wished to destroy the colony so as to gain free access to the planet's rich resources; *Seetee Ship* (1951) by Jack Williamson (writing as Will Stewart) was a direct attack on the free-enterprise system and the profit motive. In other tales, science fiction suggested that knowledge was a cooperative enterprise without boundaries of nation, race or gender. Indeed, science fiction often claimed that without this level of cooperation, science, technology, invention and development would grind to a halt.

To many, this must have sounded like scandalous communist talk, or—and which was not much better—the language of UNESCO and the UN, organizations which were routinely denounced by conservatives throughout the fifties. Worse still, the very businessmen and corporations that formed the backbone of capitalist America were held up for censure by some of the most popular science fiction writers. They were described as being backward looking, prejudiced, hypocritical and selfish. In Robert Heinlein's 'Let There be Light', included in his collection of stories *The Man who Sold the Moon* (1950), two young inventors have finally harnessed sunlight for power. They find that those committed in principle to sustaining technological advance and scientific progress actually restrict them:

> 'Free power! Riches for everybody. It's the greatest thing since the steam engine' . . . 'Decentralized cities, labor-saving machinery for everybody, luxuries, it's all possible, but I've a feeling we're staring right into a mass of trouble. Did you ever hear of 'Breakages Ltd'? It's from the preface of *Back to Methusaleh*, and is a sardonic way of describing the combined power of corporate industry to resist any change that might threaten their dividends'. 'You threaten their whole industrial set-up, son, and you're in danger right where you're sitting. What do you think happened to atomic power?' . . . 'The inventions belong to the corporation, and only those that fit into the pattern of the powers-that-be ever see light. The rest are shelved'.

The businessman—which more than any other profession in post-war America seemed to have the economic future in its hands—fares especially badly in science fiction. The corporation man, like the executive and the pin-stripe suited city slicker was, as Walter Hirsch has said, the genre's *bête noire* (Hirsch 1958). The criticism of the profession was on the whole expressed implicitly, by identifying certain negative traits in contemporary society and imagining the situation sometime in the future. Kurt Vonnegut's Jr's *Player Piano* (1952) brilliantly imagines us in a near future in which automated production has totally replaced the average worker, and in which only engineers and managers have a role to play. Ray Bradbury's *Farenheit 451* (1953) warns of book burning and of a society in which television and radio have enslaved the population, and Frederik Pohl and C M Kornbluth's *The Space Merchants* (1953), referred to in Chapter 1, sketches out a world in which business uses behavioural science to manipulate people without their knowledge or their consent.

The examples just quoted concern only one theme amongst many in science fiction, and illustrate what might be termed the approach of implied criticism— 'if this goes on. . . .' Other themes have acquired their own distinctive techniques to shock and subvert, which are sometimes less subtle and more direct. Post-holocaust tales have understandably relied for their impact on the blanket shock effect. In John Wyndham's *The Chrysalids* (1953) and Walter M Miller Jr's *A Canticle for Leibowitz* (1960), for example, vivid scenes of collapsed cities, devastation and catastrophe suggest how near to the final precipice human society now finds itself.

Space travel has, as we have seen, provided some of the richest material for science fiction speculations over many centuries. These speculations have taken at least three critical stances. The space traveller may, as in H G Wells's *The First Men in the Moon* (1901) and C S Lewis's *Out of the Silent Planet* (1938), find a paradise beyond the Earth and use that vantage point for reflections on the poverty of our meagre planet. Alternatively, travellers can meet other people who are more powerful, wiser, and less corrupt than Earthlings, and, as in Ray Bradbury's *The Martian Chronicles* (1950), discover what a pitiable chosen race we are. Or the invasion of Earth by aliens can be used directly to draw attention to the pathetic nature of life on this planet or be used to teach us a direct lesson, as happens in Wells's *War of the Worlds* (1898). Whatever technique is deployed, travel into the remotest corners of the cosmos—and beyond—always brings spaceships, spacemen, and space ideas firmly back down to Earth. Indeed, it is in so doing, figuratively or literally, that the relative (and hence criticizable) nature of the human condition can be stressed and then worked over.

Futuristic tales, holocaust and post-holocaust scenarios, space fictions and imaginary voyages: these can all be subversive, but only if they work, and whether they work or not depends, as we have suggested, on their believability. To work, they have to be anchored in reality, or at least throw light upon the human condition.

Though much science fiction was evidently conservative in its aspirations, and some attempted to import moral frames into its fantasies, the overall premise of the genre was that our established laws, customs and taboos are not eternally true constants, but are an evolving part of our planetary culture. This in itself indicated a skepticism about traditional moral positions.

The absence of a moral context in many science fiction films of the fifties is most

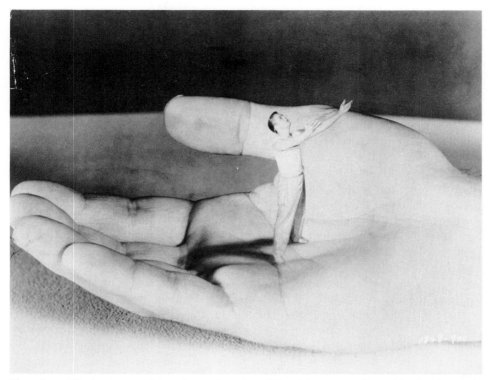

Changing attitudes to questions of morals and traditional laws are reflected .in the decreasing size of *The Incredible Shrinking Man.* In the case of this film a diminished perspective gives a new sense of awe about the cosmos.

noticeable by contrast to more recent movies and provoked a good deal of criticism at the time. It was suggested, for instance, that by failing to outline clearly the strengths and weaknesses of alien mores and ideology, and accepting instead an ethics of anything goes, film-makers were mirroring the communist rejection of 'bourgeois morality'. Some even went so far as to find in this trait the adoption of a Russian frame of values—was it not, some queried, just this aspect of science fiction which made Lenin himself such a fan of the genre? (See Highet (1953), Goldsmith (1959).) Anarchists meanwhile lauded science fiction as the most worthwhile writing of the decade and as an embodiment of many features of their world view (Pilgrim 1963). If this suggested some measure of outspoken support for science fiction's renunciation of a moral stance, the attacks upon it were certainly more visible during the fifties—and never more so than from religious quarters, which extolled such writers as C S Lewis for rendering into imaginative terms essential tenets of Christian ethics, in contrast to those who, in the words of one prominent Catholic, 'bandied morals and religion about' (McDonnell 1953).

In identifying fifties science fiction as amoral, critics as well as supporters were noting a contrast to earlier themes and treatments. With *Destination Moon*, Hollywood broke away from low-budget horror quickies and launched a series of medium-budget, solid productions. The trend towards quality and production values was increased by the introduction into science fiction of mainstream film-makers, particularly towards the end of the decade. But this was a development that took time to mature, and was always erratic. Science fiction films borrowed mercilessly for their technical devices (special effects, settings) and dramatic tones (horror, surprise), and this meant that a film from one genre could, if the budget demanded it or the director requested it, easily slide into another. The lack of a moral stance in a 'pure' science fiction film such as *Destination Moon*, which has no romantic or emotional interest at all, is not necessarily to be found in other films which drew on horror and gothic.

As science fiction became locked in a set of conventions, a sub-genre was formed of what might be termed 'science fiction gothic', in which monsters predominate. Here, there is a moral content, but it simplified and submerged—a kind of 'boy scout' morality. Take the tales of Frankenstein, for example. Harold Bloom has argued that the classic Frankenstein films (indeed those produced from 1910, the date of the first treatment, through to the Boris Karloff movies of 1931 and 1935) dwelt on the moral position of the man-made monster, just as Mary Shelley's original story had done (see Bloom (1965)). But the examples of the Frankenstein legend which were made during the fifties shifted the focus of attention away from the moral to the technical. The story no longer concerned the monster's birth, life and search for a bride (themes of the thirties treatments) but the wizardry of Dr Frankenstein himself, a man of science in his well-equipped laboratory (Formica and polished chrome) and confronted by the challenge of a little spare-part surgery. Moreover here, as in science fiction films of the fifties in general, moral disquiet (whether of the monster or his maker, alien or Earthling witness, and so forth) simply does not surface. Passions, when they do appear, are instrumental, and arrived at through ratiocination rather than being at the root of any grander ethical dilemma.

This suggests amongst other things the difficulty of drawing hard and fast distinctions between genres in this period: many science fiction films included situations and styles from fantasy and horror. But if no precise demarcation can be drawn, this is no reason to despair of ever recognizing differences. Though there are many shades of grey, it is easy to distinguish black from white. Science fiction films present the world as we know it, but transformed in a spectacular way by the application of a prodigiously advanced technology. It hinges on changes in the world as we know it, changes either caused by human beings or beyond their control—changes in the environment, in social and personal relationships, but changes brought about or bearing on scientific and technological developments. Unlike some categories of hard science fiction literature, film does not preserve the sobriety of science in its imaginings; rather it destroys it, or at least holds it up for inspection. Film renders these changes on the large scale, to suggest to viewers how experiences might change in these environments. This is perhaps the level at which film is most distinct from literature. Science fiction stories, though they also take change as their subject matter, treat it in a more subtle, abstract and reflective manner. They treat ideas, not experiences. They are more concerned with the causes of change than with its consequences.

This permits us to consider a distinction not only between science fiction films of the fifties and those which appeared earlier, but also between the genre of science fiction and that of horror and the supernatural. There is a good claim to be made, as we saw in the previous chapter, that the science fiction film of the fifties is unique in centring on science, occasionally at the expense of fiction itself, as in many of George Pal's movies. But the important role assigned to the scientist, and the shadowing of the alien or monster, stands in marked contrast to earlier treatments. In an interesting attempt to make sense of the many cross-currents, Vivian Sobchack in *Screening Space* (1987) has suggested that the horror film can be distinguished from the science fiction film in that the former takes place in a specifically and explicitly moral framework. The science fiction film, in contrast, raises issues of a socio-political character, treats technical and scientific matters, but has problems dealing with moral issues.

One consequence of the science fiction film's lack of interest in moral debate is in treatment of aliens. Dr Frankenstein's monster, circa 1931 and 1935, is sexed (indeed, the follow-up to the success of the first James Whale movie concerns the efforts of the monster to find a bride). He is sexed, and expresses his sexuality in terms which are those of the socialized human being—he wants to get married. This is a monster which is particularized: unique in his features, and humanized, finally, by his incorporation into human language. The tragedy of the film lies precisely in the way the monster functions as a kind of parodic, mirror image of his creator.

The monster of the science fiction film, by contrast, has on the whole none of these finer points. Individualism as such counted for little in science fiction terms—how could a vision that arose in the aftermath of Hiroshima? Dehumanization, rather than death or destruction, is the ultimate horror that science fiction can offer. It is the theme of man becoming unnecessary as such, of being replaced by an automaton with no loss, of machines becoming intelligent. This gives the shock-horror to themes of aliens, take-overs, transplants, robots, wars and holocausts.

The basic valuation of man as one of the hordes of animals sharing one planet, itself a tiny speck in a universe that was more than likely to be inhabited by superior intelligences—this made a focus on moral man quite impossible to sustain at the centre of the film or literature of the fifties (see Crispin (1953)). Man became defined by superficies, by the reactions he caused in other beings; it was as if in a world where everything had been thrown into doubt by science, there had occurred a disintegration between word and object, severing the connecting lines of meaning, presenting nameless things—like the unseen, unsayable, unthinkable bomb.

The bomb remains as potent a threat as it is largely because it has little adequate articulation except through the suggestion and implication of mass destruction,

Earthlings meet aliens. The anonymity of the Martians is emphasized in this still from *Flight to Mars*.

apocalypse and genocide. Aliens in science fiction films are similar in this respect. They are non-specific and generalized and, unlike Frankenstein's approachable creation, neutral. *It* Came from Outer Space, *It* Came from Beneath the Sea, *It* Conquered the World. 'It' stands in for some other, it is diffuse, it is in McConnell's phrase 'not there, but what threatens' (McConnell 1974).

Of course, the fifties screen could seldom offer unseen threats, for cinema is above all—and especially in science fiction—a visual medium; so in bridging the gap between signifier and signified, in articulating those threats, science fiction became more conventional. As it became more explicit, so it lost some of its power to shock, torment and haunt. Even so, audience reactions must have been strange. Up on the big screen, unspeakable horrors and terrifying scenarios are projected in glorious technicolor. But there is a Brechtian process (of alienation!) at work as well, since audiences cannot identify with the aliens, nor experience their experiences. With a few quite remarkable exceptions, the world is merely presented. No comment is offered, no possibility of empathy, often no density of detail, and certainly no reassuring moral point of reference. The audience are spectators. *We watch.*

We watch: science fiction offers few shocks and surprises, unlike horror and other genre films. Narrative development is central, and often pared down to barest simplicities: we watch and follow the plot. We identify not with the characters on screen but with the relationship between the situation depicted there and the situation in which we live. It is only when the distance that separates fiction on screen from fact beyond is effaced, when science fiction turns from fantasy to realism, or when characters such as the scientist are equally vivid on and off screen, that more traditional forms of identification and empathy occur. We watch: at this point moral issues can begin to be raised—about the role of the scientist, the proper functions and limitations of scientific research, the uses of technology and so forth. As we saw when we looked at the artefacts of science and the laboratory, the ethical values encoded in science fiction films are embodied in its decor and display of technical apparatus. 'Things, rather than the helpless humans, are the locus of values', Sontag has written, 'because we experience them, rather than people, as the sources of power . . . Man is naked without his artefacts' (quoted from 'The Imagination of Disaster' (1974)).

Science fiction values the instrumental over the moral. From this starting point, Sontag has proceeded to comment on the absence of horror as such in science fiction films:

> Science fiction films invite a dispassionate, aesthetic view of destruction and violence—a *technological* view, Things, objects, machinery play a major role in these films. If most fiction invents individuals, science fiction invents environments, situations.

Earlier we described the tendency of science fiction to dehumanize people, to decentre man, to see crowds as amorphous seas of expendable people. It is a genre that offers a long-distance view of man—in which people do look like ants. In so-called 'creature features' man is even further cut down: he is no longer the end of the evolutionary line, but frail and inadequate, and pitted against a hostile environment, often a bleak desert or, if not, a desolate cityscape. Nature is no nurturing mother, but the target of scientific investigations. For this it must be conceived as dead, passive and inert. Nature not only fails to provide a source of sustenance or comfort but often actively rebels: in *Beast with a Million Eyes* (1956) all the beasts of the field turn against man—horses and cows go crazy—and an alsatian dog stalks round looking for someone to kill.

Nature rebels. The death of nature, cause and product of the scientific revolution of the seventeenth century, and brought to a catastrophic conclusion with the atom bomb, has reaped a grim vengeance. Monsters from the deep, beasts awoken from their prehistoric slumber, seas and landmasses rising up in great armies—science's domination of nature is a precarious one, with a price to be paid by humankind's own degradation (Merchant 1980, Shortland 1982).

Because science either lacks total command of nature (it rebels) or destroys it (it dies), it clearly needs outside supervision. Science is not to be trusted. But because, as we have seen, science fiction depicts science in a non-moral framework, we cannot expect the weight of ethical values to act as a control on the activities of scientists anymore than it does on those of the alien. What is required is a physical presence rather than an ideological one. So it is that, in many science fiction films of the fifties, the army stands in as science's alter ego. The military is not always successful in exerting overall command, but films offer a distinctive vision of the lines of demarcation between force and reason, as well as a reflection of the real power struggles that occurred in the wake of the Manhattan project, which led up to the creation of the first atomic bombs.

National security was uppermost on the American political agenda during the golden age of science fiction—indeed during the post war period and until the late sixties—and this was on the whole a domain of policy-making sealed off from public scrutiny and influence. Decision-makers formed an elite, and a remarkably small one: from 1940 to 1967, all first and second level posts in the national security bureaucracy were held by fewer than 400 individuals who rotated through a variety of key positions. Among this elite, the traditional masculine values ruled supreme. In the Pentagon, State Department, White House and CIA, 'toughness' was described as 'the most highly prized virtue', a style of machismo that fully endorsed, indeed encouraged, violence,. As Richard J Barnet has written in his *Roots of War* (1973):

> The man who is ready to recommend using violence against foreigners, even when he is overruled, does not damage his reputation . . . but the man who recommends putting an issue to the U.N., seeking negotiations, or, horror of horrors, 'doing nothing' quickly becomes known as 'soft'. To be 'soft' . . . is to be 'irresponsible'. It means walking out of the club.

The 'club' is, amongst other things, the nuclear family. And it is in atom-bomb films particularly that the military exerts its might. In *Them!* (1954), for example, ants—grown to monstrous proportions by a genetic mutation precipitated by an atomic explosion—successfully launch a surprise attack on a US destroyer in a scene reminiscent of the raid on Pearl Harbour. Then they swarm over the country. Finally, they are cornered in a sewage system under Los Angeles and taken into custody by army detachments equipped with machine-guns and flame-throwers.

The threat of the atom bomb was the most ominous ever tackled by science fiction—a threat of whole civilizations being contaminated, the planet desolated, and the cosmos rendered unfit for habitation. It is true that atoms can bring prosperity and can hold out the promise of medical progress, as indeed was the case in the famous Warner Brothers 'biopic' *Madame Curie* (1944), in which the eponymous heroine looks with awe and hope at the phosphorescent glow of radium she has just produced in the laboratory. Ten Hollywood years later, the scene is repeated in *The Beast from 20,000 Fathoms*, but with one unequivocal difference. The glow that lights up the laboratory is the flicker of a poisoned, irradiated fish, destined to wreck havoc. Hope has become terror, and promise has been transmuted to danger. A monster has been born.

Film-makers during the fifties lost no time in discovering the power of radiation to terrorize the imagination of their audiences. Mammals, birds and reptiles were all subjected to atomic fallout, or awoken from their prehistoric sleep by an atomic explosion, to return to Earth in new, gigantic forms. *It Came From Beneath the Sea* (1955) turned out to be a vicious octopus, which was followed in unseemly succession with a veritable atomic bestiary: *The Black Scorpion* (1959), *The Deadly Mantis* (1957), *The Mole People* (1956), *The Giant Claw* (1957) and *Attack of the Crab Monsters* (1956), to name but a few. Once brought to life, the plot formula was straightforward. The beast is fearsome and seeks revenge; it crushes puny humans under foot, claw or paw. The beast destroys everything in its wake—public transport, cables, buildings, whole cities. But, like some riotous Achilles, it has a fatal weakness, an unexpected vulnerability, a romantic streak perhaps, or an insatiable sweet tooth. If the massed battalions of the combined armed forces fail to destroy the beast outright, the massed brains of the scientific establishment will study its movements, deduce its thoughts, and locate its weak spot.

If plot lines were simple, there is no denying the range of impressions film-makers were able to fashion from them. Any oversized, oversexed monster was manifestly dangerous—and the hideous result of radiation. But is atomic testing therefore to be halted, or just controlled? Can there be such a thing as 'atoms for peace'? What are the relations between force and reason? We have already seen that science fiction films from this decade manufactured an image of the scientist as 'regular guy', setting out the terms of his (or her) work and responsibilities. In the discussion which follows, we shall be looking at the relations, as depicted in science fiction films, between scientists and the military, and considering in more general terms questions of control, authority and trust.

A film like *Them!* might suggest that such problems had been resolved in favour of the military authorities; according to *Twentieth Century* (1954), its theme was unequivocally to 'Place trust in the FBI' (Anon. 1954). Science fiction incorporated with great ease much of the retinue of war films, and a lot of the visual apparatus from such films passed unnoticed, if suitably transformed, into science fiction. Dog fights, air travel, weapons, mass destruction and aerial bombardment were sure audience pleasers. The same was true for crisp, clean military personnel. In many films, overall command comes from the barrel of a gun. The army calls the shots and fires them too. *Them!* is a film which creates a particular set of anxieties and then proceeds to solve them on its own ground. Political problems are treated in crime-detection terms, with the weapon of violence being presented as the only suitable argument. Social issues are tackled with military strategy. The intelligent response is the armed intervention. The imposition of martial law is greeted with relief by the population.

But this does not mean that the military have no need of scientists or that they have fully incorporated them into the military complex. In *It Came from Beneath the Sea*, where the giant octopus, disturbed from the deep by an H-bomb, attacks San Francisco, scientists and the military begin by establishing who is in control. Having spent some time providing expertise, Professor Lesley Joyce decides she must return to her research institute to attend to other business. The officer in charge insists that she remain to help fight off the threatened attack:

> *Professor Joyce*: I am a scientist, Commander. I don't have to be reminded that your objectives are not necessarily my own.
> *Commander*: Our objectives have nothing to do with the situation . . . The Navy will see to it that you are not penalized for your absence from the Institute.
> *Professor Joyce*: Well! I feel like I'm being drafted!
> *Commander*: You are!

If this suggests the easy incorporation of science into the military machine, many other films provide contrasting scenarios. Science can certainly be harnessed for military ends; it can also lie at the very basis of modern warfare—hardly an insight in the post-Hiroshima world, it is true, but rendered more forceful in a film like *From the Earth to the Moon* (1958), since the setting is 1868. At the start of the picture, arms manufacturers from around the globe are gathered together to listen to scientist (and profiteer) Victor Barbicane's new scheme. He proposed to shoot a

The military gets to grips with science in a symbolic embrace. (From *It Came from Beneath the Sea*.)

projectile to the Moon to show the governments of the world the power of his new explosive, and hence of the new weapon that lies within his power. All immediately see the benefit of this scheme: scientific advance stokes the arms race and keeps the wheels of commerce rolling.

Such a film provides a startling account of the close and nefarious bonds between science and the military. At the opposite extreme, equally straightforward scenes from other films suggest that science may actually be a safeguard to international conflict and a guarantee of harmonious co-existence. The scientist can transmute bellicosity into peace, and the community of scientists, like the United Nations, can offer a unified vision, or offer a space in which disputes can be rationally solved. A speech in *Conquest of Space* (1955) states the case as boldly as ever it was stated. The reason for the voyage, which is diverted from the Moon to Mars, is to see whether the planet contains any useful minerals. One of the crew is a Japanese astronaut, and he suggests that the attack on Pearl Harbour and the war in the Pacific were caused by the Japanese lack of raw materials and good food. The shortage of steel forced the population to live in paper shelters and eat with chopsticks, and the poor food made them nervous, aggressive and envious. But, says Imoto, if Japan can obtain all its necessities from Mars, then it is unlikely that a war between the two countries will ever occur again.

In some cases, to be sure, the struggle between reason (symbolized by the scientist) and force (expressed by the military officer) is won by reason. *The Day the Earth Stood Still* is deeply subversive in making clear to the (privileged) viewer how mistaken the trigger-happy military authorities are in resorting to brute force to deal with an inoffensive alien. The scientist wishes to establish a dialogue with the alien, to fathom his motives, and to learn from him. Similarly, in *The Phantom from Space*, the scientist refuses to countenance destroying the monster:

> *Scientist*: The most important thing is to take him alive. If we can only understand each other, there's no telling how much science will profit. We must see it from his angle.

In truth, it is not enough to set up the categories of 'scientist' and 'military' as separate and then investigate how they are related in science fiction films of the day. After all, the fifties witnessed the development of deep tensions within the military and scientific communities over such issues as space research, secrecy, and especially over the development of new weaponry. Just as there were doves and hawks in the Pentagon and at Princeton, so too can soft and hard officers and scientists be found on screen. As Peter Biskind has suggested in his study of Hollywood films of the fifties, *Seeing is Believing* (1983), there were conservative as well as radical films depicting science and scientists. The first were inclined to let soldiers have their way over scientists, and if a dispute arose amongst scientists to favour the Tellers to the Oppenheimers. More liberal films expressed greater trust of science.

But other films lean in the other direction (rightwards, one might crudely say).

The military here is firmly in control. In *The Lost Continent* (1968), an expedition which includes scientists is led by Major Cesar Romero; in *The Deadly Mantis*, a team of soldiers and scientists is under the leadership of Colonel Craig Stevens. In other films, the military is portrayed as being almost omnipotent, able to withstand even the atom power the scientists have produced. In *War of the Worlds*, there is a doubtlessly reassuring scene in which military personnel simply brush radiation dust off their clothes after they have dropped an A-bomb on Martians; and Mickey Rooney a year later in *The Atomic Kid* (1954) is contaminated but laughs off the incident and tries to make a fast buck from it. The army witnessing a test explosion of a plutonium bomb in the Nevada desert in *The Amazing Colossal Man* sit in shallow trenches, protected only by army-issue helmets and dark goggles.

But many films did not take explicit sides, and merely commented soberly on the upheavals of the time, noting the topsy-turvy world, making full references to the events of the day—the Korean War, Russia, the A-bomb and so forth. In a world of confusion, trying to make sense was certainly difficult, as this brief snatch from *The Beast from 20,000 Fathoms* makes clear:

> *Nurse* (passing scientist a paper): What's going on in this turbulent world today then?
> *Scientist*: Oh! Death . . . and politics. The comic page is the only one that makes sense any more!

Still others offered some brilliant and sophisticated critiques of the status quo. Surprisingly enough, at a time when Hollywood proclaimed a code of production that forbade satirizing bankers, free enterprise, and when blacklists served to inhibit any criticisms of the social system, science fiction became itself an attractive escape route. It found itself able to evade the official and unofficial censorship that fell on most of the other media of communication and became an effective outlet of social protest (Shaftel 1953). It was transformed from a timid and superficial Alice in Wonderland fantasy into a voice of subversion. At a time when codes of control were rigidly enforced, how could this be so? Luck probably had some part to play. Senator Joe McCarthy himself seems to have felt that science fiction was trivial space fantasy, and not surprisingly since the last samples he had seen were just that. They were, moreover, red, white and blue moralistic and fiercely patriotic. In 1938 Flash Gordon went to Mars to defend Earth's institutions; and heroes like Bruce Gentry, Captain Video and Superman were steadfast defenders of US hegemony who used their thermal guns, cosmic vibrators and jetmobiles to defeat any menace or plot concocted to take over our world. However, the theme of patriotism did not disappear in fifties science fiction completely. *Destination Moon* describes an expedition to place the US flag on the Moon. 'By Grace of God', the leader of the expedition says upon arrival, 'and in the name of the United States, I take possession of this planet for the benefit of Mankind'.

But an alternative vision was available and was far more reassuring in its invocation of traditional values and conventions as a scheme for stability and control. This was supplied by orthodox, not to say fundamentalist, Christianity. This shifted the focus and gave human beings intellectual prominence, and reimposed an order on the chaotic world of science and technology, rather than allowing them to roll back the heavens to open up the vastness of space for an inhumanely expanded horizon. Christianity instituted purpose: no more open-ended, irresolved films; no more universes without beginning or end, no more randomness, or free sweeps through space and time. It is to this vision and more generally to the relations between religion and science fiction that we turn in the next chapter.

7

TRUE GODS, FALSE IDOLS

Religion in its traditional sense is concerned primarily with an individual's relationship with God, and only secondarily concerned with the mediation of that relationship through an organized church. In the twentieth century this essential nature of religion has become less clear cut, because people tend to see the church, temple, mosque or prayer hall as the central place in religion, as well as having their own more emotional and experiential approach to matters religious, whether or not that approach results in right living. We have seen that personal religious philosophies have led to extremes of behaviour as bad as or worse than those of the established churches, but the old religious maxim—'By their fruits shall ye know them'—holds today as well as ever.

One particular characteristic of our times is to extend the literal approach of fundamentalism (which flies in the face of tradition), which has served so well as one of the foundations of science, and to regard the Creator simply as a personal force or power, rather than the ground of all being. Such an approach is prosaic, dealing with the literal rather than the allegorical, for the natural language of the spiritual is symbolism, myth and allegory. In general, science fiction, which seldom approaches religion in its traditional sense, is the fiction of the prosaic, more concerned with narrative than poetry. As such we might expect to find little that is truly profound about religion in science fiction (and indeed we might find the promotion of atheism amongst its works) but nevertheless we can trace many and changing attitudes to religion in surprising abundance.

The decline in religiosity of western society, which has its roots well established in the eighteenth century, but which has accelerated in the twentieth, is a social, economic and cultural manifestation. Such is the power of people's natural tendency towards the spiritual that we can trace many religious substitutes, which are a means of expressing what, in another age, would have been orthodox spirituality. Science fiction, as we shall see, contains many substitutes for religion, quite apart from any tales of morality or ethical concerns which make for classic fiction in all genres.

Two of the most obvious substitutes for religious feeling in the twentieth century, and which feature in much science fiction, are the belief in science as saviour and the desire for heaven, with its angelic hosts, and a new paradise. One of the ways in which these desires are expressed is by the urge that directs us towards the stars. The journey into space is a spiritual and heavenly quest presented in technological guise. These are themes that we shall examine in more detail shortly; suffice it to add here that one sociological and personal expression of such ideas comes in response to people seeing 'unidentified' lights in the sky. Leaving aside the so far unanswered question as to what the real nature of UFOs may be (although Devereux (1982) and Vallee (1988) make sensible suggestions), from the evidence available it seems clear that 'lights in the sky' have been witnessed over the centuries and in all parts of the world and interpreted according to people's cultural, intellectual and religious frameworks or contexts.

In the Middle Ages such sights were seen as angels, God's thunderbolts or some similar manifestation of the spiritual realms, both good and bad, given as omens. Since World War II they have been perceived as technological artefacts, spaceships, time warps or aliens. Descriptions of these lights even include details of portholes, aerials, ladders, doors, fins and jets, although it is not uncommon for witnesses to add, almost as an afterthought, that the antenna turned out to be a light beam, or some such quasi-disclaimer. Carl Gustav Jung wrote a book on what such phenomena provoke in our psyches, proffering the view that lights in the sky invoke deeply entrenched archetypes in our sub-conscious minds, which have, ultimately, a religious or spiritual basis (Jung 1959). As UFOs provide rich fare in science fiction, combining the quest for heaven with the miracle of science, we shall continue this analysis in more depth when considering specific works.

If at first sight science fiction does not seem to be dealing directly with questions of religion, these reflections illustrate that in practice science fiction could almost be described as a literature searching for God, to a large extent in a world devoid of a deity that people can relate to. It does so indirectly, by analogy or by pursuing themes that have become, however deeply disguised, genuine religious substitutes. However, it is worth reiterating the idea that science fiction seldom operates directly in genuine religious tradition, except in the tradition of atheism, which is specifically endorsed in such works as Asimov's *Foundation* trilogy. Where writers invent a religious sect, or deal with specified futuristic religions, the result is far less interesting or informative than when religious matters are found by implication.

The major exception to this generality is the trilogy by C S Lewis: *Out of the Silent Planet* (1938), *Perelandra* (1943) and *That Hideous Strength* (1945). Together with his fantasy stories, the seven Narnia books, these works present a deep and rich religious mythology, showing a stark contrast to the uplifting spirituality of the celestial creatures by illustrating science to be profane, cruel and ultimately degrading. This is the science fiction of anti-science, where the glory of the heavenly hosts triumph over the evil generated by the worst aspects of the scientific mentality.

Dr Ransom, the trilogy's central character, is an academic, a philologist from Cambridge. In the first book he is abducted by Dr Weston, a scientist, and his accomplice, Mr Devine, and transported to Mars. The name Weston is aptly chosen for the villain, as he embodies all that is rotten about western society. Devine's name reflects his relationship to someone who seeks the status of divinity, but is also a play on the sixteenth century meaning of the word: foreboding. The spaceship is a standard, but only briefly described, piece of technology. Mars turns out to be a beautiful and strange mystical planet, peopled by three races but ultimately ruled by gods—the eldila. They are neither human, animal or inorganic, yet are a bit of each. They live in Deep Heaven, which we might call space, but they have no real spatial or temporal location. To them the planets of the Solar System are merely a pattern of moving points, interruptions even, of their true habitat. We on Earth have lost touch with our tellurian eldila, through our ignorance, arrogance and lost connection with nature. Dr Ransom is found to be literally on the side of the angels in this first story, whereas the scientist, Dr Weston, evolves, especially in the second novel, into the archetype of human evil.

Having said that space and science are the two most common forms of religious substitute in science fiction, it is curious here—although it is found elsewhere—that science is the protagonist. Science is opposed to religion, spirituality and the true human nature. It is the reason for our disconnectedness from nature and prevents us from putting our faith in God. It arrogantly replaces God by suggesting that *it* can provide for everyone's needs and that nature can be ignored, exploited or, worst of all, simply desecrated. Technology is underplayed by C S Lewis. He is not concerned with mechanisms. The spaceship in *Out of the Silent Planet* is merely a device for getting from A to B, and in the second book *Perelandra*, the transport is conducted by the angelic eldila, using a crystalline box. Technology is no longer needed. It rears its head somewhat more in the final part of the trilogy—*That Hideous Strength*—when a technological institution plans to re-create people in the form of robot slaves. For Lewis it is not the technology that is interesting but the battle between the sacred and the profane. He simply borrows the clothes of more mainstream science fiction to present an image of the evil product of total profanity. This battle is a subject that he has written about as a polemical argument in *The Abolition of Man* (1943), and which he here presents in fictional form. The sacred, he is saying, needs no technology, for it springs from a truly other realm of which our material world is but a lower level in the hierarchical creation, that has been described elsewhere, from Plato to Dante. It is in this model of reality that Lewis is most at home.

Just as in the Narnia stories, where explicit religious symbolism is found in the characterization, in Lewis's science fiction we are led into the worlds of mythology. The final battle in *That Hideous Strength* involves the planetary gods, mythical figures such as Merlin and Arthur Pendragon, fairies, nature spirits and angels. The richness of the spiritual realms, Lewis is telling us, far exceeds the paltry vision of the scientist, whose conception, even of space and time, for all their complexities, is prosaic and mundane. If we look into our hearts we can find the truth, and from there it can flourish. The spiritual nature of man, and of nature too, is the ultimate reality and to turn to science can only be to turn away from truth. The Lewis trilogy is a powerful science fiction argument against our scientific culture, and it is an argument conducted on religious grounds.

The extent to which Lewis was influenced by Fritz Lang's film masterpiece *Metropolis* (1926) is not known. He does not refer to it in his correspondence.

Certainly Lang's vision is similar to that of Lewis, although Lang does not describe the spiritual realms. *Metropolis* is also against science and technology, but on sociological grounds, even though it contains a specific religious argument. The main thrust of the film's narrative is that love can transcend class, even in a society where a small, powerful technological elite completely dominates the workforce. This is essentially a film with a social and political rather than a religious message. It projects into a futuristic world, with expressionist architecture, faintly reminiscent of Manhattan, where technology is quite explicitly shown to be the means of mass oppression. The concepts are quite extraordinary for its time, showing both a fascination with a scientific future, and a horror of it.

The exploited masses work under ground, in a nether world (shades of Heironymous Bosch) which combines a feeling of the catacombs with a mechanistic technology that Chaplin showed to perfection in *Modern Times* (1936). The Orwellian masses have become slaves to the machine, working until they drop to keep the mechanisms that support the city going. Nothing is permitted to prevent the machines from functioning, for this would spoil the elegant lives of the elite. Humans, however, are frail and so we are introduced to the 'mad scientist', who, as in Lewis's third novel, is designing robots to take over from the unreliable human work force. Such is the mix of emotion and intellect in this character that his prototype robot is a woman, a machine perfect in all respects, an object of beauty to love and marvel at. She/it stands in contrast to the heroine, Maria, whose love for the factory owner's son eventually brings unity after the slave-workers successfully revolt. Maria is the archetypal virgin. Tender and determined, her faith surmounts the evil and degradation around her. She leads the workers in Christian worship in an outlawed subterranean church, and it is through the imagery of the crosses in her chapel that the religious theme enters the story.

Lang presents a contrast between the hedonistic smugness of the technologists who form the ruling class and a spiritually uplifting church, which survives suppression and outrage because it promotes higher values and offers hope, salvation, beauty and truth, personified by Maria. The message being presented here is that science supports the rulers, indeed makes the rulers powerful. It is a message understood and reiterated by many writers and science critics through the last five decades. Lang goes further—he also equates the power–science duality as a religious theme in itself. Science is the new religion in Lang's *Metropolis*, and it is worshipped explicitly in the huge, thinking head that acts as a focus for the elite's worship. It is a religion based on idolatry, and the robot goddess, deliberately modelled on Maria, is a second and undermining personification of this idolatry. Technology as a religion, this story tells us, is idolatrous and diabolic. 'By their fruits shall ye know them' applies to both sides of the religious contrast painted in this story. The true religion is on the side of goodness and the idolatrous pseudo-religion is expressed through hatred and oppression. When science and technology become religious substitutes all havoc breaks out.

The apparent political implications of *Metropolis* are, on reflection, more subtle than at first appears. Criticism of communism, slave labour and the suppressed

Christian symbolism in a world of scientific materialism. The underground, illicit church of the workers in *Metropolis*.

religion, which looks like the Soviet system at its worst, is modified by the equally strong criticism of capitalism. Both systems prize their technologies and both are attacked in this film, as is the scientific foundation of their power. Analysis of Lang's political intent leads one to the conclusion that although his political stance may have been naive, advocating a general principle of brotherly love, which was found in much German thought of the 1930s, there is, nevertheless, a sharpness to his attack on the technological base of all systems. This attack is found, too, in the writing of the Čapek brothers (Karel Čapek's *R.U.R.* (1921) gave birth to the word *robot*) and in the work of Zemiatin.

Where C S Lewis and Fritz Lang condemn science as a religious substitute as intrinsically evil, leading people away from their own humanity and God's will, what we find in Arthur C Clarke's *2001* (1968), and even more explicitly in Stanley Kubrick's 1968 film on which the novel was based, is science and space leading us progressively to the Almighty. Here science and technology are idolized as man's stepping stone to immortality; 'What men once dreamed of we can now actually do'. *2001*, in the first instance, is a eulogy to science and technology. The waltz of the spacecraft is a sheer, exuberant psalm to the goodness and beauty of space and its

Science worships a false idol. The elite in *Metropolis* are more concerned with pleasure than principle and their form of worship is presented as idolatrous.

technology. The ape sequence at the start of the film immediately places the subsequent material into an evolutionary schema. Ape–man–superman is the line of progress and the dissolving image of the tossed bone weapon becoming a spaceship marks the transition from one important evolutionary stage to another.

2001, however, does not present a Darwinian account of evolution. This is very much engineered evolution, with the mysterious black obelisk appearing to our ape ancestors and stimulating their consciousness. It is found again on the Moon, tempting us to take the next stage and travel to the next beacon, the obelisk at Jupiter. This is evolution guided by a superior, mysterious and external agency. Is this god an angel, a devil or an alien—a mysterious but powerful intellect from dimensions outside space and time? What else are all these but different forms of presenting God? The story even suggests that our innate religious feelings, our desire for the heavens and their creator, comes from the fact (within the context of the story) that we have evolved knowing deeply within our subconscious minds that our creator (genetic manipulator) comes from above, from the skies, in a mysterious, unfathomable fashion.

Once we have evolved enough to get to the Moon, which requires a certain level of technical ingenuity, then the next obelisk (or is it the same one?) awaits. Right away we have an implication that this mysterious intelligence puts technological skill high on its list of what it is to progress—hence the hymn to space travel. This is not altogether surprising because we are dealing with an intelligence that is itself technological. The obelisk is manufactured in the sense that it is definitely an artefact, although way beyond anything we can comprehend. It is, in the technological sense, the equivalent to the wardrobe in the Narnia stories, an entry point to higher dimensions and another reality. Yet the obelisk is not an old wardrobe, it is a piece of technology.

The yearning for space as a heaven substitute is at its most elegant in *2001*. The film was made in the heady days of the late sixties, when the first Moon landing was in near prospect (1969) and the first pictures of Earth from space took on an enlightening quality, heralding in a new age. The journey to Jupiter, as the next step in the story signifies, is, like all journeys in stories, an analogy for progress, the search for truth, the spiritual quest. There is something 'out there' which is both mysterious and awesome. It is calling us and we must seek it out, because it will transform us, maybe make us immortal. Heaven awaits, but instead of turning inwards to find it, as the mystics teach, this heaven is outside and we must travel into space to find it.

The goal is reached, although many of those involved in the quest were not worthy of the task and perished along the way. For the lone survivor, Bowman, a destiny awaits. As he approaches the obelisk he utters the last words that will be heard from him: 'My God! It's full of stars'. Bowman is propelled by the obelisk through space–time into the unreality of a strange bedroom, beyond the light of the stars. Is this a kind of purgatory or is it heaven itself? Then death comes and with it a rebirth as the star child, symbolic of both the Christ figure and the human embryo. This is the birth of a new man, a superman, as well as the reincarnation of Bowman. Maybe it is all hallucination, or is it that the mysterious and unseen intelligence is too much for the human mind to bear, and so can only be presented in confusions? We are to it what the apes were to us. We have still to evolve to that supreme point, and the transformation of Bowman is, maybe, the next stage in evolutionary manipulation. The philosophy behind this is very much akin to that of Teilhard de Chardin, and not close to that of Darwin. The obelisk is leading us towards de Chardin's Omega Point—another God substitute, although for the heterodox Chardin the substitute seems very much like the real thing.

This is a remarkable parable, both a paean to technology and a religious analogy that is very direct in its approach. Many people believe today that heaven/space is our destiny, but this belief was perhaps more widespread at the time of the Apollo Moon landings and the release of *2001* in the cinema. As a film *2001* is an extraordinary achievement—beautiful, enigmatic and conveying something of the awesome quality that such a religious theme should inspire. The sequel, *2010* (1984), completes any mystery left over from the first film. The obelisk really is the door to heaven and Bowman, as a ghost, transforming in appearance from young man to old man to star child, visits those he loved like an angel with the message that something is about to happen. When asked what his reply is, 'Something wonderful'. It turns out to be the conversion of Jupiter into a star by a vast injection of mass by the multiplication of obelisks. Political conflict on Earth is brought to a halt by this new 'Star of Bethlehem' and finally we are left with the image of the primaeval forests on one of Jupiter's moons, that has now become a new planet. Standing in the undergrowth is an obelisk. God moves in mysterious ways and none more so than in Arthur C Clarke's two stories. Whilst suggesting God is a superior intelligence (which is theologically quite sound) neither the sequel nor the original story goes so far as to suggest that God is, quite simply, an alien. The idea lies beneath the surface of this sophisticated and intelligent work, but it is quite explicit in the even more popular work of Stephen Spielberg.

ET—The Extraterrestial (1982) is a quite deliberate presentation of the God-as-alien motif. The alien is found (i.e. born) as a creature from above and discovered in a shed (stable). It is only recognized by those without power or authority (shepherds), in this case children. Adults are seen, for the most part, from a child's perspective—mostly from the waist down—and the children are clearly the dispossessed in relation to adults. The alien is persecuted by 'occupying' forces (Romans/police), dies and is reborn. The risen alien, dressed in a white robe from head to foot, reveals its love for the children by showing its throbbing (sacred) heart. It ascends back into heaven. The Christ story is explicit in *ET*. Despite the fun, the chases and the American cuteness, this is science fiction with a religious substitute theme and there is something slightly off-key about it. This is not because of the supernatural powers, even though these seem to be exercised mostly with children's electrical toys and ultimately with bicycles, nor is it the technological undertones that enable ET to 'phone home'. The trouble with ET is basically that this God substitute is really rather ugly, crude and quite obviously plastic!

Of course, it can be argued that most people will enjoy this film simply as entertainment and that those who notice, or care about, its obvious form as religious substitute will realize how poor that substitute is. However, if the God-as-alien presentation is not directly perceived then it will work subliminally on the audience, effectively subverting their religious views (if they have any) and telling them that God is simply something like this plastic doll. At best the message will permeate the minds of the audience that the Christ story, with all its power and uncompromising challenge, is nothing more than whimsy. The danger of a film like *ET* lies not in its possible blasphemy but in its subverting imagery and in its effect of undermining traditional religious thought.

There is a danger in explicitness that the cinema of science fiction has continually to cope with. Unlike the written word, which leaves so much to the reader's imagination, science fiction on the screen either has to suspend its credibility with awful models and other effects, or to spend vast fortunes to make them plausible, or it loses its grip on the material by leaving things unseen. In *2001* both money and the unseen are found together. Whatever aliens are around remain undescribed, and are all the better for it. Films like the Star Wars series have used special effects and models to great effect, while other science fiction films, like *The Day of the*

Surrounded by a halo of light, draped in a white robe, *ET—The Extraterrestrial* is presented in an iconography of Christian symbolism.

Triffids (1963), are ruined by quite awful, even risible monsters. Occasionally, in a film like *It Came From Outer Space* (1953), the aliens are only half depicted, but so indistinctly that the imagination fills in the missing information with notable effect. With *ET*, despite the expense of the animated model, the alien-Christ is still a model. Spielberg gives it lovable, large eyes, which as Disney knew always induces sympathy, but this does not compensate for its plastic appearance. As a God image ET runs counter to the whole history of religious iconography. This is anti-icon, and the theology of such imagery has yet to be explored.

Like Christ, ET is anti-authoritarian, but whereas Christ's criticism was directed to the ossified authority of the Temple, the film's approach is to contrast the naive alien with a menacing, masked and sinister para-military authority. The world of adults is seen to be unnerving and socially manipulated by forces which remain anonymous. The same forces are seen at work in Spielberg's *Close Encounters of the Third Kind* (1977). Here the police and military forces manipulate even their own collaborators in order to mask their true intent. Although this turns out to be quite innocent, unlike the threats to ET, Spielberg creates the same unsettling feeling of menace with his authority figures. It is as if he believes there are official conspiracies around, even when he is party to them.

If *ET* is the Christ story in whimsical science fiction, *Close Encounters* gives us the full God-as-alien spectacle. Here we have the super-technology of lights in the sky—the small UFOs encountered earlier in the movie, whizzing around and disrupting electric toys again, are just the winged chariots, messengers of the gods. The Mother Ship (although the imagery which surrounds it is entirely technological and hardly smacks of femininity) is apocalyptic in scale. It finally appears like an Old Testament God of Wrath, manipulating the clouds before it, creating lightning storms and generally being heralded by mighty omens. Spielberg is surely trying to tell us that this is God!

The awe this craft inspires in the participants of the story, and through them to the audience, is powerful in its feel and import. When the hero of the story steps into the craft we discover that it contains a god-like centre, full of might and power, and throbbing with noise and light. This is the holy of holies, which in the Judaic tradition became charged with electricity when at its most hallowed. Our hero has crossed the threshold from this world into the next, never to be returned, to be

The final sacrifice—or sacrament—as Richard Dreyfuss is led into the Mother Ship by a group of aliens, his arms spread out as if being crucified. Stephen Spielberg makes further use of Christian imagery in *Close Encounters of the Third Kind.*

transported into the heavens for the new life for which he has been prepared. He is attended by the angel hosts, but once again these angels are ugly. Although only half seen, and immersed in the light of the craft, they are not beautiful like angels are meant to be, and can only give rise to speculation as to what heavenly realm they have come from. One clue to the orientation of Spielberg's religious substitute imagery is the location for this ultimate encounter. The place for the descent, transfiguration and re-ascent was selected by the aliens themselves. It is in Wyoming, and it is called the Devil's Tower.

The symbolism of this place of encounter is intriguing. It is a location of startling physical appearance, a dramatic volcanic plug reaching up out of the plains, which, in the story, becomes implanted in the minds of those who have encountered the UFOs. Devil's Tower is sealed off by the authorities, who pretend it has become poisoned, and who kill off farm animals to make the pretense look real. There is something sinister about the location and the tactics by which its isolation is enforced. This is despite the apparent innocuousness of the encounter itself. What morality lies behind this political and military manoeuvring? Why is it condoned by the supposedly superior beings in the spacecraft?

The authorities arrange for a special team (who look like astronauts) to go with the aliens in exchange for those who, over the years, have been abducted. All this suggests a collusion between the authorities and the aliens and leaves open the question of what the aliens did to those they had removed from Earth. We never hear their debriefing, and only glimpse a child looking happy to be back with his mother, but also slightly sad to have left the alien craft. The menace turns to a mood of joy when the exchange takes place. Everyone looks happy. The hero offers himself up to join the team to go on the craft and is half carried by the aliens, into the light of the ship's belly, with his arms outstretched like Christ on the cross. It is his offering, freely made, to leave this world for the next. It is presented partly as a sacrifice, partly as a fulfilment of his destiny.

The preparation of the other 'sacrificial' disciples is completed with the only explicit religious moment of the film, when they join in prayers conducted by a parson, who blesses their future as they leave for the UFO. It is an oddly placed scene, and jarring in its blandness, when compared with the awe-inspiring poetry devoted to the aliens. Church, Spielberg is saying, is pretty mundane stuff compared with the real gods.

God as alien is a common theme in science fiction. If Spielberg is blatant in his adulation of the mysteries of space and its inhabitants, he is not alone in his use of imagery. *The Day the Earth Stood Still* (1951) also presents a Christ-like man of peace from the heavens with supernatural powers, who is rejected, as is his message, and who is looked upon with scorn and doubt. Like Spielberg's aliens his power over electricity is commanding. His guardian angel is a robot, who can both harm and heal. The alien brings hope to those who have faith. He assumes the name of Carpenter—Christ's trade—and is betrayed, killed and resurrected to ascend back into the heavens. Like ET, he is the Christ figure as alien. An English (Anglo—angel) actor, Michael Rennie, was cast in the part, and, an alien to the

United States, was different enough to be an outsider. Another English actor, David Bowie, appeared as another alien/angel figure in *The Man Who Fell To Earth* (1976). Here the alien angel is tragically debased and betrayed by the humanity he had turned to for salvation.

There is no real ambiguity about Mr Carpenter. He is seen as nothing but good and kind, bringing a message of peace, even if he is single-minded and resolute. His determination, however, seems healthy enough and his 'good news' is genuine, even though it is backed up with veiled threats of destruction. As has been argued earlier in this book, this film is placed in a strong political and social context, but the alien as God here is Christ-like, with strong undercurrents of Jehovah. In *Invasion of the Body Snatchers* (1956) it is the unseen alien that is worrying. The alien might be a god but it might also be a devil.

There are too many cases to consider where the things from space are hostile. From H G Wells' *War of the Worlds* (1898) to *Alien* (1979) we have encountered too many horrors out there. Maybe this is where a religious theme is supplanted by a purely psychological one. There is a strong argument for treating outer space as a metaphor for inner, subconscious space and the monsters as representations of our own suppressed fears and complexes. This idea was itself explored in the excellent science fiction film *Forbidden Planet* (1956), which was loosely based on Shakespeare's *The Tempest*. There the subconscious became actualized as monsters from the Id, rather like the power prevalent in *Solaris* (1971). It is not surprising if our reaction to a visitation from up there is one of suspicion, caution and a readiness for self-defence. The God-like visitor is probably the exception, and if not deserving our full attention as an object of veneration, it could at least be given some respect. In *The Day the Earth Stood Still* it is only the scientists who have time to listen, for they are the good guys, in contrast to the stupid politicians and suspicious military. But if our reaction to the alien is not over generous it could be because we suspect that a technological god is not the real thing. The scientists, who listen to Mr Carpenter, are prepared to take him as a saviour, but then they already have a God substitute—technology. We are not asked to worship Mr Carpenter as we are the aliens in *Close Encounters*, simply to bear in mind his message. The UFO aliens at Devil's Tower have no message for us at all, they are simply there to be worshipped. If we are dubious about that it could be because of the place they chose to land. They might be telling us something unpleasant about themselves, or maybe the place name is a Freudian slip in the mind of Stephen Spielberg.

Another face of God in science fiction, also associated with space and sometimes with aliens, is the spiritual credence given to the idea of a superior intelligence, either disembodied or suggested as being around but with a thin or veiled material presence. The eldila of C S Lewis's novels take this form, but are explicitly angels, and a similar approach is found in *2001*. The idea crops up as the central notion in Fred Hoyle's *The Black Cloud* (1957), although here the intelligence is given a body, albeit a very tenuous one. Hoyle makes crucial the theme of a superior mind which is so awesome that anyone who comes into contact with it dies of neural pathway overload. Even the main character of the novel, a Hoyle self-portrait, can tune into

The mask of the god *Zardoz*, from the film of the same name.

this positivistic god for only a little longer than any other mortal. It takes at least a great physicist to listen to the messages of the Almighty, and we need to evolve smarter scientists to reap the full benefit such a voice from heaven can impart.

The relationship between science and religion is ambiguous in John Boorman's film *Zardoz* (1973). The god Zardoz is the idol to whom the 'eternals' of the story pay homage. These 'eternals' are the elite whose immortality comes about purely from the scientific life support system of the Vortex, where they live. Theirs is no true immortality but an artificial one—a technological means of living for ever. This community of 'eternals' have but one desire, death, which is their only means of finding freedom and salvation. The Outland people, a brutish mob, also seek freedom, that of equality. They want to share the science of the Vortex, which they ironically see as a path to freedom, whereas the elite find science the unfreeing bond that ties them to their fate worse than death—everlasting life. It is a confused film, philosophically and dramatically, but it is another instance of science and technology being substituted for religion.

The unseen aliens in *2001* appear to have the power over life and death, which is a necessary qualification for a deity. The unseen intelligence that stands for God in Andrei Tarkovsky's *Solaris* (1971), which was based on Stanislaw Lem's outstanding novel, takes this concept even further. This is an intelligence which is possibly located on the mysterious planet, it may be the planet itself, or it may be a god of the mind that uses the planet as a psychological mechanism to induce hallucination to create a mystical desire in those who fall under its spell. This is a presence that lurks, that has supernatural powers, that can conjure up and make material whatever you wish and yet leaves you dissatisfied. This is an intelligent god who makes you turn your desires and fears into realities, so that you might reject what you thought you wanted and in the end want only God—to be united with the Creator totally. All the other fears and desires were themselves only substitutes for God in any case. Finding this truth is like returning home, that what you sought elsewhere was really with you all the time. This is a theme that Tarkovsky has re-worked in several films, including the strange, haunting science fiction piece *Stalker* (1979). This is actually orthodox religious teaching par excellence.

The difference between the two films lies in *2001*'s preoccupation with the technological and the theological predominance of *Solaris*. A comparison of the two approaches reinforces the idea that the Creator is not to be found in hardware, even if both stories invoke a feeling of awe about the cosmos. The contrast between the two movies also distinguishes the religious backgrounds of the USSR and USA. America is the land of fundamentalism and sectarianism, where the technological is closely allied to religion. The God promoted on the television channels and the worship of the space race are both tied together within the American mind. The Soviet Union rejects religion formally, and *Solaris* maintains a facade of this, but the Russian people are deeply pious and their religion is solidly orthodox and traditional. This comes across very strongly in Tarkovsky's work.

The physical universe, symbolized in the night sky, is a motif that is frequently used as a background for religious speculation or cosmic wonderings. This setting often brings us closer to a genuine religious feeling than when the conquest of space acts as a religious substitute. In *The Incredible Shrinking Man* the final scene shows the hero, now reduced to about the size of an ant and still shrinking, musing on the infinities of space, as he escapes from the cellar to the outside world. Suddenly we are taken from purgatory, which has been faced with courage and ingenuity, where despair has been overcome and hope is a real possibility, into the freedom that the heavens offer. The infinite wisdom of the Creator is extolled—one who can offer meaning and peace, even when man has shrunk to nothing. In contrast to the Almighty, man is but a grain of dust (in this case literally). After the adventure we have a sermon, a perspective on the miracle of life and the hereafter.

The heavens may bring its terrors, as in *It Came From Outer Space*, but the terrors may be only of the unknown or the unfamiliar. Our approach to these fears is all important. If we are to face them fairly, see that the alien is not so unlike us, and regard the unknown as we would like to be regarded, then salvation is possible. The losers are the bigots and the bullies, those whose gut reaction is to shoot first. To kill the unknown, the alien, is to remain alienated; to embrace what you fear is to be redeemed. This is good psychology, right social action and satisfactory theology. At the start of *It Came From Outer Space* an amateur astronomer, surveying the stars through his telescope and extolling their wonder, spies a meteorite that turns out to be a crashing spaceship. This discovery leads to the encounter with the aliens and a battle for good over evil. As it happens the evil comes not, as it first seemed, from the aliens, but from the prejudice of the local people. When all has been righted and the aliens have left in peace we are treated to another homily on the heavens and of a brighter future. Here there is more than a hint of the religious substitute, because the astronomer surmises that one day we too will be up there among the stars, exploring new worlds. The implication of the goodness of that future is both a reference to the life hereafter and to the goodness (i.e. the religious substitute) of space exploration.

Not all science fiction equates that seeking after knowledge with good. In *Conquest of Space* (1955) science and religion become opposed again. Progress is not equated with goodness. Man was born on Earth, it is argued, and the Bible never mentions him going to other planets. But despite this fundamentalist warning, this fear of the physically and spiritually unknown, space is conquered, or at least it is explored. The religious qualms are expressed not so much as a criticism of scientific progress, rather as a condemnation of a religion that would turn back the clock and hold us in its own form of oppression. This notion is seen fleetingly in *Close Encounters*, when the brief prayer meeting makes the religion look small-minded and un-important. Here science is the new creed, and the old religion must make way for it.

When Worlds Collide (1951) is a parable along the lines of the story of the flood. The latter day Noah and his team of experts take off for a new world when poor old Earth is about to be destroyed by the rogue planet Zyra. They take in two of every kind (but all American) and the inheritance of (western) culture. The Bible is pre-eminent amongst the books of the spacecraft's library, but no priest is to be seen. Whilst the media on the threatened Earth tell us that 'All the world prays', bringing all people closer to God than ever before, the team of evacuees leave for the heavens, not to be closer to their Maker, but to build a new life on another planet. They have expelled the evil in their midst (by shooting those who stood in their way) and eventually they land in their new Eden. The scientist in charge claims the priestly role both by his behaviour and by his protection of orthodoxy, as well as by publicly leading the prayers, giving thanks for their saved lives. The planet turns out to look tranquil but it also contains mysterious structures that look suspiciously like churches. Let's hope the natives are friendly!

In *From the Earth to the Moon* (1958) there is no great shared enthusiasm for the desecration of space, and the spaceship's captain, Dr Nichols, shoulders the burden of guilt for the mission. To him it is sacrilege. When the rocket runs into trouble he blurts out 'It will not be my hand that kills us, it will be the avenging hand of God!' Here is the scientist disclaiming the outcome of his own profession. He symbolizes the science critic, seen more aptly against the background of the arms race or speaking out about the problems of ecological catastrophe. Nichols' argument against the space race is less social or political in import, but is primarily a plea for religious fundamentalism.

As in C S Lewis's *That Hideous Strength*, it is made clear in *Conquest of Space*

Journeying to the stars is a blasphemous act for the doubting astronaut in *Conquest of Space*.

that there are arguments to suggest that some knowledge just should not be known. As Mars looms up on the ship's screen the Captain is told 'It is getting larger'. 'Yes' he replies, 'the planet and the blasphemy'. Mary Shelley, in the original *Frankenstein* (1831), introduced into science fiction the question of the ethics of scientific progress and knowledge. Although the idea that some things should not be known is as old as religious thought, it was in the stories about creating monsters that the examination of this idea was brought to science fiction. The crude expression of it is that there are things which only God has the right to do and know. In Genesis the argument is more sophisticated. You should not eat from the tree of knowledge of good and evil, lest you become like gods. The usurping of the place of God, the identifying of humanity as the be all and end all, is the original sin of the fall, not knowledge in itself. It is what knowledge does to he who knows that becomes the sin. The urge to explore is another form of religious substitute and one that frequently expresses the desire to overcome a fear of the unknown. The problem in exploring new ground, however, is exemplified, often enough, by its results.

The blind quest, without any thought for the consequences, means playing God, and this is the real sin. In *The Fly* (1958) the scientist who discovers how to disintegrate and then reintegrate matter after its transportation (just as Willy Wonka does with his television in Roald Dahl's *Charlie and the Chocolate Factory* (1964)) is obsessed with the technique rather than its context. Admittedly he discusses its useful application, but ignores any idea of its problems. When the experiment throws up the unexpected and he becomes half man and half fly, he destroys the whole experiment so that it cannot be duplicated. He has unleashed an evil that demands his own death as expiation for his errant behaviour. We are left with a dehumanizing technology that is best destroyed. It is the Faustian legend in science fiction, and it is retold in countless stories.

The dehumanization of people through technological oppression and the rule of the tyrant (who may be the scientist or his allies) leads to another religious substitute theme; that of politics and power as God. This is seen in *Metropolis* where the power figures and their idols are eventually overthrown, and more dramatically, and realistically, in Orwell's *1984* (1949). Big Brother is a god substitute. The opium of the masses is provided by the State, and although the main purpose of the narrative is to make social and political comment, there is also a religious flavour to the idea of the triumph of the human spirit over the forces of oppression.

Although Asimov's robots obey his three Laws of Robotics and become the perfect slaves and helpers of mankind, the original idea of robots in Čapek's play *R.U.R.* was that technological man, dehumanized, turned ultimately into a piece of technology, is no longer man at all. The human spirit is what makes us human (the word *human* comes from the Greek *pneuma*, meaning spirit and breath), and to turn people into machines is to take away our possible salvation and final union with God. So the science fiction of the machine man, even when it is being advocated at its strongest, is still an assault on religious principles. Asimov rejects the spiritual

Innocence in the face of evil. A child offers an apple to the stone monster in the powerful silent classic film *The Golem* (1920).

nature of man and glories in his technological progress. For him, as for many people in the twentieth century, religion is a thing of the past, a primitive hangover to evolve away from or to surmount. Yet underlying such thoughts and ambitions there remains a belief in humanity that continually triumphs over the forces that oppress us and try to lead us away from our humanity. The robot symbolizes that state, and it is a fallen state from which we need to be redeemed. Even in the science fiction of atheism there remains a core of religious feeling, even when the author is trying to reject it.

The approaching millenium brings with it a renewal of religious feeling, whilst rejecting the restraints of the organized churches. Our new heroes are still religious but are harder to place in a religious spectrum than was the case in the past. We are now in an age of mysticism and science fiction reflects this by its move away from technology and its redirection towards the ecological, the awareness of the living cosmos and the unity of all things. The scientist as hero is replaced by the sage and the fiction is more readily labelled as fantasy than as science fiction. It may be that the glamour of science has worn rather thin, and the need to explore unknown territories can be better satisfied through altered states of consciousness and other pathways to the gods.

The idea of the scientist as god, worshipped both directly and through his inventions, is no longer as appealing as it was. The price for that status was always high and too often paid for by his death or through a transformation that led to his redemption. That tale has been told again and again in science fiction, and so we no longer feel the need or the obligation to worship at the altar of science as we used to try to do, even when warned of the dangers of that path by the science fiction writers themselves. The religious themes in science fiction are also often the retelling of older, more perennial morality tales. In this chapter there has been no real exploration of morality as such, nor of the ethical issues that are the substance of many science fiction stories. Morals and ethics are not the exclusive province of religion and could really deserve their own separate exploration. Of course, a book like James Blish's *A Case of Conscience* (1958), where a Jesuit priest is one of the main characters and has to partake in a moral issue concerning the destruction of a planet, should not be ignored, but such stories do not fit precisely into this discussion. Where the ethical and moral tales become interesting is where the substitutes for the religious iconography of science and space lead to a deeper questioning of our attitudes to what might be called 'the big questions'. Science fiction gives no clearer an answer than any other form of art. That it deals with questions of theology is itself an assurance of the broad scope and universal comment and concern of its authors. Rather than appear as a simple form of entertainment, science fiction actually explores the deep issues of our times as widely as does any fiction, and raises questions that are of perennial concern.

8

GREEN NIGHTMARES

Since the advent of the nuclear bomb the notion of a world-wide holocaust has haunted the minds of millions. Mankind found itself with unimaginable destructive power and the idea of a world devastated by nuclear war, leaving genetically damaged inheritors of an Earth barely habitable, was explored in the imaginations of many. Apocalyptic thought has, arguably, been focused by the coming millenium, a traditional time of foreboding, and such ideas have been enhanced by the growing awareness of the devastation to the planet Earth through pollution, overpopulation and ecological damage.

Rachel Carson's influential book on pollution, *Silent Spring* (1962), opened up a new and growing awareness of how civilization was tipping the scales of natural balance in the environment. Thirty years later global ecological catastrophe is probably a greater fear than the terror of nuclear warfare. It has become realized that natural disasters can have greater global impact than even a superpowers' nuclear war. The effects may not be so immediately dramatic, but the consequences could be more devastating. Earthquakes, volcanic eruptions and impacts from bodies from space all release much more energy than the hydrogen bomb. Earthquake energy can be measured in thousands or tens of thousand of megatons of TNT: explosive energy and impact by a minor asteroid would be measured in millions of megatons, dwarfing the world's stockpile of nuclear weapons in comparison.

Science fiction writers have delighted and terrified us over the years with a wide range of apocalyptic stories, inspiring feelings of awe in us by the magnificent terrors of the phenomena and yet reassuring us (in most cases) that we would survive, as we identify with the surviving heroes of the tale. However, it is with the consequences of disturbing the balance of nature that this chapter is most concerned. Such changes may be triggered by warfare or by natural disaster, such as volcanic eruption, collision with an astronomical body or even by instabilities in the output from the Sun, as was the situation in J G Ballard's novel *The Drowned World* (1962). Natural balance may be altered by pollution, radioactive waste or fall-out, plague, genetic engineering mutations or a thousand and one other means. The science fiction writer explores the results, more often than not, to portray the resulting altered world and how people cope with the situation.

Nineteen fifties science fiction was most preoccupied with the effects of nuclear holocaust and the possible consequences of radiation damage. Nevil Shute's *On the*

Beach (1957) was the seminal post-nuclear war novel, made into a powerful film by Stanley Kramer (1959). *Five* (1951) and *The Day the World Ended* (1956) presented similar themes but the more insidious and unexpected results of radioactive contamination both reflected contemporary concern and fear of what was an unseen but possibly present danger and explored mankind's altered relationship with society and nature through this genre. *The Incredible Shrinking Man* grew smaller because he had become engulfed in a radioactive mist and his decreasing size was a metaphor for the altered human perspective in the face of a nuclear world. *Them!* and *It Came from Beneath the Sea* (1955) both presented wildlife contaminated and mutated by radioactivity, becoming monsters seeking revenge against the human race who had messed up the natural order. Even the amusing *Mr Drake's Duck* (1951) laid radioactive eggs. In the case of *Tarantula* (1955) the mutation is the result of scientific experiment going wrong with the result that a town is subjected to an invasion of man-eating spiders, a theme to be returned to in later years as concern grew about the scientific (and implied unnatural) interference with the natural.

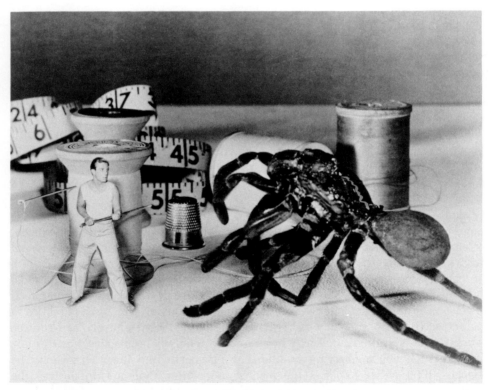

Man versus nature in a battle where the odds have been changed by altered scales. *The Incredible Shrinking Man* continually questioned the human perspective on life as the hero got smaller and smaller.

The fifties and sixties also presented us with a whole range of disaster movies. *When Worlds Collide* has the threat in the form of a collision with another star, scientifically very implausible, although the later *Meteor* (1979) had more credibility, perhaps encouraged by the renewed interest in the 1908 Tunguska meteorite fall in Siberia. Larry Niven and Jerry Pournelle's *Lucifer's Hammer* (1977) had a comet impact in the ocean, causing torrential rains, the blocking of sunlight and a new ice age. Earthquakes are the theme in *The Night the World Exploded* (1957), whilst the sixties film *Crack in the World* (1965) was similar except in so far that the disaster was caused by an experiment to tap the heat from the Earth's core as an energy source, an experiment that went wrong, cracking up the Earth's crust. In *The Day the Earth Caught Fire* (1961) atomic bomb tests have unwittingly altered the Earth's orbit, taking it closer to the Sun. There is a strong suggestion in these stories that science is going too far in its manipulation of nature with results that are both unacceptable and terrifying.

One of the most influential science fiction books of the fifties was John Wyndham's *The Day of the Triffids* (1951). Here was an ecological theme presented with extraordinary clarity. The breeding of triffids, strange hybrid plants that could move about, had been done for the exploitation of their high oil yield. An innocent enough endeavour! When a natural disaster strikes and radiation from comet showers causes almost universal blindness, the triffids, no longer attended, proliferate and become dominant, not least because they have highly poisonous and powerful tendrils. Interference with nature, in this story, is shown to be a disturbing affair, with consequences that could not be reasonably foretold.

The sixties film of the same story (1963), unsatisfactory largely because of the artificiality of the simulated plants, was untypical of the decade. The main concern of the sixties was still largely focused around the nuclear issue. *Dr Strangelove* (1963) was the apocalyptic movie of the decade, its black humour echoing the irrationality of the arms race.

'Gentlemen, you can't fight in here, this is the war room', snaps the President to the bawling Russian Ambassador and a US General. The sinister Doctor Strangelove himself (reputedly based on a cross between Henry Kissinger and Edward Teller) epitomized the cold rationality of the science behind the military technology. Joseph Losey's film *The Dammed* (1961) and the film of *Lord of the Flies* (1963) by Peter Brook both illustrated the deep concern about radiation effects in the aftermath of nuclear war, a concern that was only enhanced by the Cuban missile crisis. The appearance of *The Satan Bug* (1965) heralded a new worry, chemical and germ warfare already being practiced, at least in part, in Vietnam.

The science fiction films of the seventies expressed very clearly a concern that had risen to a general awareness, that the interference with nature could lead to a retaliation from the natural world. This was not a new understanding but one that had become widespread, perhaps for the first time. *Frogs* (1972), *Squirm* (1976), *The Kingdom of the Spiders* (1977) and *Day of the Animals* (1977) all featured strong themes of the revenge of nature. The last of these told of a natural backlash as the ozone layer was destroyed by man's stupidity and greed, a theme that would have been even more topical at the end of the eighties. *No Blade of Grass* (1970) presented a world where a virus had destroyed the world's crops and *Prophecy* (1979) took up the topical worry about mercury poisoning. A new theme was that of overpopulation. *Zero Population Growth* (1971) presented a world where to have children was a criminal act and *Soylent Green* (1973) painted a picture of overcrowded cities, food shortage and mass euthanasia. Where science fiction writers were voicing ecological concern in the fifties, the movies were re-discovering this theme in the seventies and to great effect.

In the seventies the science fiction films still vaunted fears of nuclear catastrophe. *Quintet* (1979), an intelligent movie made by Robert Aldrich, is set in a post-nuclear holocaust world where the nuclear winter, a new ice age, has come about. Radioactive fall-out is the attributed cause, rather than the dust that blocks out sunlight in the scientific version of this scenario. It is a bleak film, where the only unfrozen emotions are between the hero and his girlfriend, until she dies unable to give birth to their child. The implication of the story is that everyone else will die too. *The China Syndrome* (1979) focuses on the dangers of civil nuclear power, rather than weaponry and the early eighties film *Chain Reaction* (1980) takes up this theme with nuclear waste as the protagonist.

The eighties have seen little science fiction dealing with the environment, tending to resort more to fantasy and escapism, certainly as far as the cinema is concerned. *Testament* (1983) stands out as the one enterprise dealing with the aftermath of nuclear conflagration and *Warning Sign* (1985), although not much more than a tough thriller, is set in a laboratory where germ warfare viruses get spilled, causing those infected to turn into psychotic killers. In *Quiet Earth* (1985) the hero finds himself alone in a world where scientific experiment has virtually resulted in the total annihilation of the populace, but in a decade of growing concern with the balance of nature, the rise of the green politics movement and a genuine threat to our planet from increased pollution, the degradation of the ozone layer and AIDS it is surprising that the cinema has not responded to people's fears and concerns in the way it had done over the three previous decades.

The turn to fantasy may be a reaction to the dawning realization that the futurologists of the sixties and the science fiction writers, whose doom-laden forecasts and stories had excited a public who had 'never had it so good' and who liked the thrill of imagining the hardships of a world gone wrong, might have been right. The scale of global ecological crisis grows daily, with the greenhouse effect warming the planet, the ozone layer depleting, the rain forests being destroyed to a frightening extent, seas polluted by oil spills, radioactive waste and effluent: all these and many more are brought home by the media daily. The consequences and implications of these tragedies are, it can be argued, too awful to think about and so, when the media offers irrelevant escapism, it is turned to avidly.

Ecology is the science of the interdependence of all the factors in an environment, animals, plants, the soil, water, air and man. Ecological disturbance occurs when an extraneous, changed or otherwise radical factor occurs in the ecosystem. An imbalance results in a chain reaction, affecting the state of the environment as a whole and leading to a shift in the relative size of the populations that inhabit that environment. The ecosystem may be a meadow, affected by the removal of a hedge or by pesticide spraying, it may be the planet as a whole. Technological developments, increasing urbanization, growing population size and the subsequent usage of finite natural resources all lead to global changes that can affect the health, well-being and indeed the survival of the human race.

The misnamed 'greenhouse effect' (greenhouses have more to do with shielding from wind, restricting air movement and allowing convection currents to warm the space than trapping infrared rays of sunlight) brings about a warming of the atmosphere as carbon dioxide builds up in the atmosphere, the result of burning fossil fuels, and which acts as a heat shield around the Earth. A global temperature increase of only two degrees Celsius would result in major changes of climate, the creation of new deserts and land loss due to the thermal expansion of the oceans, helped marginally by the melting of the ice caps. The degradation of the ozone layer, allowing ultraviolet radiation from the Sun to reach the Earth's surface, will add to this effect as well as increasing the incidence of skin cancer. Holidays in the sun may not be such a good idea in the future.

Deforestation not only removes the habitat of many species but it also reduces the major source of carbon dioxide absorbers, the trees, and leaves the soil open to the processes of erosion. Removal of the tree cover may expose land that can be used in the short term for quick cash crops but is likely to result in further desertification. The major disruption to the myriad species that live in the forests can lead to plagues of insects and crop-destroying infections, apart from the reduced diversity of species. Chemical fertilizers, used to replenish the nutrients in what was fertile land pollutes the water of the region, destroys the balance between aquatic plant and fish life and eventually fouls the oceans.

Nature exists in a beautiful natural balance. Any perturbation to that balance is redressed. Major disruptions result in a larger scale adjustment but the scale of interference can push the system into chaos. The natural environment seeks its own harmony, not mankind's idea of what is useful, and the consequences of the interference we are now collectively inflicting on the world ecosystem will not be kind to us. We cannot murder the planet but we can cause genocide. Life on Earth will continue, whatever we do, for millenia long after our civilization has faded into the dust of history.

The ecological crisis has developed as a result of the growth of population and of technology. The natural environment can cope with degrees of pollution but, as waste products such as carbon dioxide, nitrates and radioactive waste accumulate, permeating the air, the oceans and the food chains, so the biosphere as a whole gets overloaded. The development of technology parallels the growth of scientific knowledge and the adoption by advanced civilizations of the scientific approach to the world. The attitude we have towards nature has shifted from one of deep understanding to one of manipulation and sentimentality. The harm we do to the planet is a symptom of an underlying malaise. The growth of science and technology has come at the expense of wisdom, and this is linked to the spiritual decline that has been traced by Acquaviva in *The Decline of the Sacred in Industrial Society* (1979) or presented by Theodore Roszac in *Where The Wasteland Ends* (1973). The spiritual decline of the West has resulted in a desacralization of the natural order, of using nature for short-term gain, for having power (or the illusion of power) over natural forces. Civilization after all means living in cities and city folk rapidly lose touch with the organic cycles of the countryside. Apart from what has been until recently a small number of ecologists warning us of our folly the group of people whose ecological message has reached the widest number of people has been the science fiction writers. Even when the ecology of their stories has not been to the forefront, in many cases the message that nature can and will strike back has been presented forcefully.

The most famous of all monster movies is undoubtedly *King Kong* (1933). Although it falls naturally into the category of stories where the past is explored by finding a lost world, where evolution has remained partially stuck in the era where dinosaurs roamed the Earth, hence allowing drama to take place in another kind of world, the film also has a powerful ecological message. The story concerns a film-maker and entrepreneur who wants to record the legendary gorilla on film if he can. He is taken to Kong's island by a tough boat crew, who would sooner shoot to kill than to photograph. The capture of Kong is an unexpected bonus and the colossus is taken back to New York to be displayed as a wonder to a paying audience. Nature, the film is portraying, is there to be exploited, made inharmonious, gaped at. However, the film does not have its sympathy with this viewpoint. King Kong has our sympathy throughout. When the poor creature escapes to protect the one human it loves, the glamorous Fay Wray, our heart goes out to him. He is a mis-understood creature, just as nature itself is misunderstood, and the shooting down of Kong by aircraft using machine guns is a metaphor for the destruction of the natural order in a wider sense. Ultimately *King Kong* is a sad film. It provokes feelings of anger and frustration at the ignorance of those who exploit the beast and in the animal's death we are led to feel a genuine grief. Kong's innocence, and ours, have been destroyed.

The notion of natural revenge, of nature striking back, is a common theme in science fiction. The change in humankind that allows the triffids to take advantage of their new-found situation is a prime example, illustrating Darwin's theory of natural selection. The mobility of the plants, their ability to withdraw their roots, lurch on and dig in again, combined with their deadly, blinding sting, combines notions of ecology and natural selection with the writer's ability to take what are in this case carnivorous plants like the Venus Fly-trap, and exaggerate and extend into what might be possible. Even the smallest details in Wyndham's book show how nature redresses the balance. It doesn't take long, for example, for grasses to push their way up through concrete, reclaiming for their own the land that had been sterilized to make way for roads. The fragility of our civilization is well illustrated here.

The disaster movies pick up this theme and delight as well as scare audiences with natural disasters, giving the special effects team full reign and usually showing

The most famous 'monster' of them all! *King Kong* roars out his challenge to those who would hunt him down.

The devastation by natural forces can be spectacular. The poster design for *Krakatoa, East of Java* shows the famous volcano erupting.

the indomitable human spirit bearing out against the odds. *Krakatoa, East of Java* (1967) showed the full impact of volcanic eruption, as the illustration, taken from the poster design, shows. This was drama based on the 1883 eruption, which released hundreds of megatons of energy and threw so much debris into the sky that sunsets around the world were spectacular for years. The film reconstructed the eruption and its accompanying tidal wave, which in reality measurably travelled right around the globe seven times. Three hundred towns and villages were destroyed and over forty thousand people lost their lives. Greater fears, however, have been triggered in audiences by earthquakes.

Hollywood is built on the San Andreas fault, the geological crack in the Earth's crust that runs from northern California down to Los Angeles. It forms part of the Pacific coastal region of North and South America where crustal plates abut and which comprises part of the 'ring of fire' that runs right round the Pacific, and on which Krakatoa also lies. In 1906 San Francisco was reduced to rubble by an earthquake associated with this instability. It was a disaster compounded by fire, which raged through the broken buildings, and exacerbated by civil disorder and looting. It was reconstructed in the film *San Francisco* (1936) although this was not a disaster movie as such. Mark Robson's megabuck epic *Earthquake* (1974) was designed to be the ultimate in such movies. The idea behind the film was to show the destruction of Los Angeles. The fictional quake would wreak havoc on the skyscrapers in the city centre and rupture the dam holding back the waters of the Hollywood reservoir. Los Angeles would perish from crashing buildings, the consequent fire and finally by devastating flood. To add to the spectacle the producers used a new sound system: Sensurround. This took the form of an additional sound track on the film containing noises at the lower end of human audibility and below it. The electronic vibrations amplified through the theatre replicated the feel of an earthquake just as the pictures illustrated it. The advertising warned of the effect and included a disclaimer for responsiblity to audience reactions. Certainly *Earthquake* was spectacular but as with most such films the human emotions, plot development and audience involvement with the characters suffers from the swamping special effects.

In *Volcano* (1953), however, the smouldering and pent-up emotions of the central character, played by the sensual and sultry Anna Magnani, are paralleled dramatically by the anticipated eruption of the volcano of the title. When the film reaches its climax and the heroine puts to right the human disaster she has caused, she fittingly dies in the volcanic explosion, which has been orchestrated to highlight the metaphor between human emotion and its counterpart in nature. John Ford's uncharacteristic and romantic film *The Hurricane* (1937) also matched dramatic emotion with, in this case, the wind. So effective were the special effects, showing a Pacific island being devastated by the fury of the gale, that the effects director, James Basevi, won an Academy award for the hurricane sequence.

Natural disasters certainly illustrate the fragility of our way of life and refocus our perspective on the awesomeness of natural power, which dwarfs that of our most destructive weaponry. Although characters in disaster movies are frequently heard to say lines such as 'this is the wrath of God' or 'We are being punished for our iniquities' such disasters are seldom seen as nature hitting back against mankind. One way in which this really does occur is to be seen in the development of new bacterial strains that are resistant to the anti-biotics that have been developed by the pharmaceutical industry. Attempts to become immune to their effects are subverted by the mutation of new strains. This is nature striking back, not by vicious intent, but merely by the desire for survival of all species. This situation is seen most commonly with influenza but AIDS has brought the reality of something much worse to everyone's mind. It provokes fears within us of something truly deadly, akin to the great plagues of the middle ages, and carries with it overtones of concern and repressed emotions concerning sexuality and the social mores of our society.

George Stewart wrote about a deadly plague, killing off the people, in his book *Earth Abides* (1949). The few survivors found themselves in a world that had rid itself of an even more threatening disease, human beings. Nature had found a way to restore itself to health by removing the human infection, by reducing the offending population to levels of complete harmlessness. In *The Andromeda Strain* (1970) the virus that threatens life on Earth comes from space. A satellite crashes on a New Mexico village, infecting the small populace with its alien and deadly potency. Most of the story is set in the laboratory where the virus is contained and eventually destroyed. The author, Michael Crichton, is himself a medic and so one of the group of science fiction writers who have been trained in

A deadly virus strikes down the population of a New Mexican village. Scientists in protective clothing inspect the scale of the problem and prepare to deal with *The Andromeda Strain*.

science. Robert Wise's film, faithfully based on Crichton's book, was seen by its director as being more science fact than fiction. It stresses the irony of the scientists having to trust their technology in the face of defeating an enemy whose character changes, through mutation, during the battle. Can their computer model outwit the natural process of mutation in order to find the Achilles heel of the virus? That is where the tension lies in the film, which has a remarkably authentic ring to its science presentation.

Another scientist with science fiction credits to his name, mentioned several times in earlier chapters, is Sir Fred Hoyle. His non-fiction works include *Diseases From Space* (1979), in which he argues that viruses are extra-terrestial by nature and that life at a primitive level is abundant in space. Provocative though his ideas may be, and they have been received with considerable hostility and derision by a majority of the scientific community, they nevertheless provide a link between fact and fiction that may yet turn out to illustrate how science fiction can anticipate science fact to come. It has been suggested that microbes from space would not be suitably adapted to prey on terrestial organisms. In Harry Harrison's *Plague From Space* (1965), when germs do infect humans it turns out to be because they have been designed to do so by aliens as a form of interstellar germ warfare. This is an idea where science fiction mirrors the fears people may have about what we are doing, or might do, to ourselves. We humans can behave as badly as hostile aliens inevitably are meant to.

The ecological lessons to be taught by the science fiction movies of the seventies are implicit rather than generally stated. A film like *The China Syndrome* is exceptional here, because it presents a case against the development and use of civil nuclear power for electricity generation. In the USA its warning message was enhanced by the Three Mile Island incident, which resulted in a curtailment of the nuclear power programme. *Silkwood* (1983) was science fact, dealing with the problems of big business interests and their incompatibility with safety concerns over nuclear power. It is more an industrial thriller as it portrays the events leading to the death of Karen Silkwood, just as she was about to deliver evidence of malpractice in the nuclear industry, in which she worked, to the authorities. These are both stories with a political and incidentally an ecological statement to make and neither deals with the notion of nature hitting back, rather than with our endangering the natural order. When nature is shown to retaliate it does so, at least in science fiction, in a gross fashion.

The Beginning of the End (1957) is typical of its type. A secret laboratory has discovered, however implausibly, that the fertilization of crops by a radioactive agent results in huge plant growth, increasing yields enormously, but at a price. The insect life that also feeds off the crops is affected and in this story the particular insects are grasshoppers. They, like the plants, grow huge and then turn nasty. The result is ten foot high man-eating grasshoppers ravaging Illinois and terrorizing the population of Chicago. In *Squirm* the creatures are worms and their transformation has come about through electricity, when a power cable collapses near an isolated farmhouse. The science fiction element and the revenge of nature motif are little more than excuses to drum up a horror story. Nobody would like to turn on a tap and, instead of water, have flesh-eating worms emerge, as happens to the local sheriff, it is enough to put you off your spaghetti.

Chemical sprays are what trigger the changed behaviour of tarantulas in *Kingdom of the Spiders*. Less of a shock-horror story, this film has much more style and scientific realism as nature bites back. The images of the dead in the Arizona town, where the film is set, who have been the victims of the spiders' group behaviour, lying in the dust cocooned in webs to provide later nourishment for the arachnids, is quite chilling, especially as the spiders have won. Here tampering with nature has a direct consequence as mankind's ecological niche is usurped.

With *Willard* (1971) the menace is from rats, while *Empire of the Ants* (1977) brings back radioactive waste as the trigger and excuse to mix actors and highly magnified film clips of the innocent little ant. *Frogs* benefits from not trying to explain why swamp creatures have turned nasty. All the audience, and the characters in the story, know is that the animals are on the warpath. Neither are the creatures magnified, but remain their true selves physically, and become even more menacing thereby. The idea of a heroine being lured by butterflies into a swamp, where leeches are waiting to attack her is pretty frightening stuff and the ecological theme, that nature might do to us what we do to it, is all the more dramatic from its inference, rather than from a trumped up and pseudoscientific explanation.

Nature striking back is perhaps the oldest theme in science fiction, pre-dating what we might regard as science fiction proper. Edgar Allan Poe's short story 'The Masque of the Red Death' makes use of a plague as a device to sweep the world free of cruelty, excesses and moral depravity. When the human race neglects its role of stewardship and responsibility, then death and destruction may well follow. Throughout the history not only of literature, but philosophy and religious thought, apocalypse is the natural consequence of collective sin. From the Garden of Eden, where man is banished from an earthly paradise, through Sodom and Gomorrah to the world of science fiction the redress of harmony with God and the created world is found through disaster, plague, famine and destruction. What the science of ecology has done is to formalize such notions in respect to the environment and open the world of true science fiction to expand and comment on what seems like natural law.

The apocalyptic nature of nuclear war, its absurdities and terrors, are taken to their extreme in *Dr Strangelove*, its despair illustrated in *Quintet*, and its surrealistic logic presented in *The Bed-Sitting Room* (1969), which is one of the rare examples of science fiction originally presented as a stage play, whose zanyness in the face of nuclear holocaust comes from the writing of Spike Milligan and John Antrobus. All these stories point to the folly of a technologically sophisticated yet emotionally and spiritually retarded civilization. It is folly that science fiction can portray all too clearly. The consequences of nuclear warfare give rise to pictures of menace and deprivation, yet there is sometimes a degree of ambivalence to the story lines.

John Sturges' film *The Satan Bug* (1965), based on Alistair MacLean's book, for

Nature strikes back. Ernest Borgnine finds that rats are not the friendly and cooperative little creatures that he thought they were. (From *Willard.*)

example, is a science fiction thriller concerned with germ warfare and the development of a deadly virus in the laboratory. The plot concerns the murder of one of the scientists working on the dreadful bug in the desert research laboratory. The frequency with which the desert features in science fiction stories and films suggests that there is a significance in the aridity and barrenness of such places as a metaphor for the state of neglect technological society has got into; an ecological comment in itself. The investigation that follows the murder takes us into a routine thriller plot. A deranged millionaire is the villain. It turns out that he has stolen the virus in order to hold the US government to ransom in an attempt to force them to call a halt to the germ warfare programme. He plans to threaten Los Angeles with the virus, evidence enough of his paranoia, reinforcing the message that the germ warfare programme itself is a sane approach to international relations. Of course, the heroes catch him and recover the lethal flasks of virus in the nick of time, thereby saving Los Angeles and the germ warfare programme. Very patriotic stuff, but hardly a lesson in ecological sense, or even of political wisdom.

John Christopher's novel *The Death of Grass* (1956), filmed by Cornel Wilde as *No Blade of Grass*, is an account of one family's struggle to survive in a world where a 'satan' bug has destroyed most of the planet's greenery, particularly the edible crops. This dystopia presents a picture of a fragile society that has gone beyond what we would hope would be the future and, as in the disaster stories, shows our civilization as being very friable. Death, violence and a parched and arid land face the hero and heroine, demonstrating the impact that human interference with nature can give rise to. Wyndham's *The Chrysalids* (1955) paints a similar picture of a threatened society, just as *The Day of the Triffids* did. Mankind's folly, resulting in a holocaust, has left a remnant people in whom genetic mutation gives rise to frequent abnormalities in the children born to an irradiated populace. Such abominations must be stamped out. Wyndham's humanity, however, uses this situation to portray the essence of our species as lying in its soul and not in outward appearance. A plea for tolerence is made with the passion and heart that Tod Browning brought to his strange but sympathetic movie, *Freaks* (1932).

Whatever the cause of the holocaust, life still goes on and many science fiction works are concerned with the reconstruction of society from the ruins of civilization and from a ruined earth. Both *The Death of Grass* and *The Chrysalids* fall into this category. *Planet of the Apes*, as described in Chapter 3, is set in a post-holocaust future with life re-adapted in a new ecological order. Many of these works are concerned with the condemnation of the technology that led to the devastation and the sterile thinking behind it. In Edmund Cooper's *The Cloud Walker* (1973) religion is focused on the Luddite Church, whose officers wreak havoc on anyone caught re-inventing technology. Science fiction writers, however, generally do not want to do away with the very stuff of their genre, so all too often the technology is re-discovered and reconstituted. Often this is presented as a comment that mankind just cannot leave well alone and to point up the maxim that 'you can't stop progress'. In the most sensitive and intelligent of these stories, *A Canticle for Leibowitz* (1960) by Walter Miller, the monks who preserve pre-holocaust knowledge after the

Third World War do so to enable civilization to be rebuilt. When all is restored another nuclear war is the inevitable result. Science fiction writers, in general, do not seek a truly alternative pathway for the human race, but then their concern is, by definition, that of dealing with *homo scientificus*.

One mechanism to refurbish a devastated planet proposed in science fiction is to preserve life forms in space. Douglas Trumbull's film *Silent Running* (1971) presents us with a veritable Garden of Eden in an enormous space station. It is tended mainly by robots, as well as human gardeners, and it contains forests, pastures, etc. The intention of its construction is to provide the stock for replenishing the Earth, should catastrophe strike and the need arise. When the head gardener is ordered to destroy this world in orbit he rebels, killing the remaining humans and releasing the modules with their different ecological environments so that they can drift away into space, tended by their automatons and possibly to provide a place for a new evolution, a new life. It is a sad film in certain ways, commenting on human folly and destructiveness, yet optimistic that nature can be preserved, even with a human gardener, and could permit a new and presumably better race of humans to grow in its midst. The garden of *Silent Running* provides a true ecological message of our duty to preserve and tend the creation around us.

Ecological catastrophe is just one of the ways in which life processes are interfered with by science and its attendant culture. Experiments that go wrong lead to such ends as is presented in *Tarantula*, giant spiders that set upon their human masters or the monsters that radioactive pollution gives rise to, as in *Them!* and *It Came from Beneath the Sea*. Although not apparently of an ecological nature, stories such as *Frankenstein* are nevertheless about our attitudes to nature and man's desire to control it. Dr Frankenstein's desire to be able to create life, to act like God, is the urge that leads to technological domination and the destruction of the ecosystem. There is a marked difference in human behaviour resulting from, on the one hand, a belief that wisdom is embodied in nature and, on the other hand, the belief that nature can be improved upon by human ingenuity. The first case leads to behaviour moulded to maintaining balance with the environment, the second leads to chemical fertilizers, crop sprays, deforestation and the rest. It is the mentality of the second case that is embodied in the traditional 'mad scientist', who eventually can be found paralleled in science as an institution and that plays a crucial role at the heart of modern civilization. The creation of a living being will inevitably be a monster for two reasons. Firstly because it has no ecological niche, it is *de facto* artificial, born out of nature not of it, and secondly, because, if it has been made autonomous, it will not be under the control of its maker. It is the notion of control that is so important here. It is monstrous not to be under control, to be uncontrolled, out of 'my' control. The philosophy of science is very much concerned with the 'conquest of nature', bringing the wilderness, the wild, under control.

It would be unfair, however, to always blame the scientists. It is the accompanying mentality and the desire to control that is the dominant factor and it is just as likely, if not more so, to be found in politicians, industrialists and, of course, the military The BBC television series *Doomwatch*, in the early seventies, was based around a small government unit of scientists, set up to investigate pollution and environmental problems. The series looked at problems of an ecological nature and all too often in the stories it was the scientists who found themselves up against the military, political chicanery and industrial might, all protecting their interests against the common good. It was a highly responsible, informative as well as entertaining series.

The adverse effects of pollution itemized in weekly doses of *Doomwatch* are added together in John Brunner's powerful and depressing novel *The Sheep Look Up* (1972). Although like *Doomwatch* a work of fiction Brunner's novel is a story based closely on what was at the time observable in the USA, but written with everything much worse. It catalogues the effects of pollution of air and water, and the degeneration of the land, and continues relentlessly, leaving the subcontinent aflame as nothing can be done to reverse the damage inflicted, unwittingly, on nature. Bob Shaw's *Shadow of Heaven* (1969) parallels this account, where herbicides have created dustbowls across the globe and people only have synthetic food to eat, supplemented by some seafoods. Part of the problem in both these cases are the people.

One of the themes that has dominated ecological thinking, particularly in regard to the usage of natural resources, has been that of overpopulation. Many of the excesses of civilization's impact on the environment come simply from the sheer size and growing number of people. With the world population doubling in shorter and shorter intervals of time science fiction writers have had to find ways of dealing with them. Four basic solutions can be traced out. In one catastrophe, natural or inflicted by man, reduces the population dramatically. The holocaust scenarios deal mostly with that case. Another solution is to feed and house the growing numbers in increasingly artificial environments, like the massive tower blocks, each containing a million people, in Robert Silverberg's *The World Inside* (1971). The third solution is to shift people into space colonies, finding new worlds to inhabit and, finally, the least common approach in science fiction is that of birth control.

In *Zero Population Growth* (1971), set in the twenty first century, the world has become so overpopulated and polluted that the authorities make it illegal for any children to be born for a period of thirty years. Naturally the film focuses its attention on a couple who defy the ban and have a child, but the picture painted of a future with too many people has a certain realism to it. The most powerful film on such a topic, however, is *Soylent Green* (1973), based on Harry Harrison's *Make Room! Make Room!* (1966). The story is set in New York in the year 2022. Overpopulation and insufficient food (steak, fruit and vegetables change hands illegally for huge sums on the black market) and the populace are fed mainly by the synthetic products of the Soylent Corporation. The hero, a tough policeman, is investigating a murder which leads him to discover the grim secret of Soylent Green, one of the food products. His one friend is an aging scientist, played touchingly by Edward G Robinson, who died shortly after the film was shot. In the story he has a last meal with Charlton Heston, the cop hero. His ecological message, which was Robinson's final screen speech, says that the problems engulfing everyone are due to human contamination and rottenness, but, as Robinson says, 'the world is beautiful'. The next day he goes to the centre where all older people are compelled to attend. It is a euthanasia centre. Before death the people are shown pictures of the way the world was, when it was still beautiful, and in such a state of peacefulness they die. The happy memories invoked at the moment of death are little more than a guise for genocide. The corpses are then transferred to the Soylent factory and processed into feedstuff. Soylent Green is people.

The message from all these stories is one of gloom. The picture painted of the future is not one to relish. Violence erupts as overcrowding brings out aggression in people. The fight for survival, even when institutionalized, as in *Soylent Green*, leads to further oppression and even to cannibalism. Death, deformity and destruction accompany the apocalyptic horsemen. This, say the science fiction writers, is where we are heading. If you don't like it, change things now.

The film with one of the strongest ecological warnings is one that is not always considered to be science fiction. It is Alfred Hitchcock's *The Birds* (1963). The message is not explicitly stated, for this is a subtle and complex work, open to interpretation at many levels. Loosely based on a short story by Daphne du Maurier, *The Birds* tells the ominous story of altered bird behaviour. The birds become aggressive, attacking humans at will in mass, coordinated assaults. At first this strange behaviour intrudes only occasionally into what appears to be a love story, with girl chasing boy. Then it develops into a more complex examination of repressed emotions, psychological tensions and barely concealed family relationship difficulties. The birds become ever more intrusive, reflecting by their aggression the release of pent-up emotions in the human story, but eventually taking over the picture so that the film becomes one of survival against the devastation the birds are wreaking on the human race.

In a central and key scene, just after the birds have attacked the children at the little school in the Californian fishing village in which the action is set, a conversation takes place in the local bar between most of the characters in the story. An eccentric English lady, an amateur ornithologist, is protesting that the account of the attack cannot be true.

'Birds are not aggressive creatures, Miss,' she says, 'they bring beauty into the world. It is mankind, rather, who insists upon making it difficult for life to exist upon this planet.'

A drunken Irishman, sitting at a nearby table, and frequently quoting the Bible, intersperses with, 'I tell you, this is the end of the world!' To which the lady replies, 'I rather think that a few birds are hardly going to bring about the end of the world.'

When she asks the heroine of the story why the birds attacked the school children the response is, 'To kill them'. 'Why?', 'I don't know'. Neither do the audience. No explanations are offered in the film and no happy ending either. After the last major assault on the main characters, locked and boarded up in their house, the birds in their millions just sit around and wait. The hero, with his mother, sister and the woman he loves all get into their car to try to get back to the city and, hopefully, away from the birds. They drive off cautiously and the film ends with

Ecological warnings are there in plenty. Alfred Hitchcock's *The Birds* shows the horror and the illogic of ecological disaster. These illustrations come from the director's storyboard for the film.

complete uncertainty about the future. All we see are birds everywhere.

Returning to the old lady in the bar, we find that as her disbelief starts to dispel, so her anxiety increases.

'I have never known birds of different species to flock together. The very concept is unimaginable. Why, if that happened we wouldn't have a chance. How could we possibly hope to fight them.' She witnesses an attack and is terrified. This is nature striking back in earnest. The collective stupidity of humankind has resulted in an awful retribution. As the Irishman said 'It is the end of the world.'

BIBLIOGRAPHY

Abernethy, Francis E, 'The Case For and Against Science Fiction' *Clearing House* **34** 474–7 (1960)

Acquaviva, S S, *The Decline of the Sacred in Industrial Society* (Blackwell, 1979)

Aldiss, Brian, with David Wingrove, *Trillion Year Spree* (1986)

Allott, K, *Jules Verne* (Cresset, 1940)

Amis, Kingsley, *New Maps of Hell* (1960)

Anderson, Poul, *Tau Zero* (1970)

——Technic Civilization Series, see *The Long Night* (1983)

Anon, 'Them' *Twentieth Century* **106** 197–8 (1954)

Apuleius, *The Golden Ass* (Penguin, 1964)

Arendt, Hannah, *The Origin of Totalitarianism* (Harcourt, 1951)

Ash, Brian (ed), *The Visual Encyclopaedia of Science Fiction* (Triune, 1978)

Asimov, Isaac, *Asimov on Science Fiction* (1984)

——The original Foundation Trilogy consists of *Foundation* (1951), *Foundation and Empire* (1952) and *Second Foundation* (1953)

——The Foundation Series consists of the trilogy listed above together with: *Foundation's Edge* (1982), *Foundation and Earth* (1986) and *Prelude to Foundation* (1988)

—— *The Gods Themselves* (1972)

—— *I, Robot* (1950)

——'It's Such a Beautiful Day' (1954) in *Nightfall and Other Stories* (1969)

——'The Martian Way' (1952) in *The Martian Way* (1955)

——'Trends' (1939) in *The Early Asimov* vol 1 (1972)

Asimov, Isaac, Greenberg, Martin Harry and Waugh, Charles G (eds), *The Science Fictional Solar System* (1980)

Bachelard, Gaston, *La Poetique de l'espace* (PUF, 1967)

Bainbridge, William Sims, *Dimensions of Science Fiction* (Harvard, 1986)

Bainbridge, William Sims and Dalziel, Murray, 'The Shape of Science Fiction as Perceived by the Fans' *Science Fiction Studies* **5** pt 2 (1978)

Ballard J G, *The Drowned World* (1962)

Barnet, Richard J, *Roots of War* (Penguin, 1973)

Barron, A S, 'Why do Scientists Read Science Fiction?' *Bulletin of Atomic Scientists* **13** 62–5 (1957)

Barthes, Roland, *Mythologies* (Jonathan Cape, 1974)

Bates, Harry, 'Farewell to the Master' in *Astounding SF* October (1940)

Baxter, John, *Science Fiction in the Cinema* (Tantivy Press, 1979)

Beasy, Phillip, 'God, Space and C.S. Lewis' *Commonweal,* **1** 421–3 (1958)

Bell, Daniel, *The End of Ideology* (Harvard, 1960)

Benford, Gregory, 'The Secret of sf is Awe' *New Scientist* 23/30 December, 765–7 (1976)

—— 'The Designers' Universe' *Amazing Stories* November 109–16 (1988)

—— *Timescape* (1980)

Benford, Gregory and Brin, David, *Heart of the Comet* (1987)

Bester, Alfred, *The Stars My Destination* (1956)

Biskind, Peter, *Seeing is Believing: How Hollywood Taught us to Stop Worrying and Love the Fifties* (Pantheon, 1983)

Blish, James, *A Case of Conscience* (1958)

—— The Cities of Flight Tetralogy consists of *They Shall Have Stars* (1956), *A Life for the Stars* (1962), *Earthman Go Home* (1955) and *A Clash of Cymbals* (1958)

—— *The Quincunx of Time* (1973)

Bloom, Harold, 'Afterword' see *Frankenstein* Mary Shelley (Signet, 1965)

Bogdanovich, Peter, 'Interview with Don Siegel' *Movie* **15** 1–14 (1968)

Bradbury, Ray, *The Martian Chronicles* (1950)

—— *Farenheit 451* (1953)

Brancourt, Guy, 'Interview with Don Siegel' (1970) see *Focus on the Science Fiction Film* ed William Johnson (Prentice-Hall, 1972)

Brin, David, the Tales of Uplift consist of *Sundiver* (1980), *Startide Rising* (1985) and *The Uplift War* (1987)

—— 'Tank Farm Dynamo' in *The River of Time* (1986)

Brookeman, Christopher, *American Culture and Society since the 1930s* (Macmillan, 1984)

Browning, H E and Sorrell, R A, 'Cinemas and Cinema-going in Great Britain', *Journal of the Royal Statistical Society* **A117** pt 2 133–70 (1954)

Brunner, John, *The Sheep Look Up* (1972)

Burnham, John C, 'American Medicine's Golden Age: What Happened to It?' *Science* **215** 1474–9 (1982)

Burroughs, Edgar Rice, *A Princess of Mars* (1917)

Caidan, Martin, *Cyborg* (1972)

Čapek, Karel, *R.U.R.* (1921)

Carson, Rachel, *Silent Spring* (Penguin, 1965)

Carter, Everett, 'The Scientist as the New Hero' in *Howells and the Age of Reason* (Lippincot, 1950)

Cartmill, Cleve, 'Deadline' in *Astounding SF* March (1944)

Christopher, John, *The Death of Grass* (1956)

Clarke, Arthur C, *Childhood's End* (1953)

—— *The City and the Stars* (1956)

—— *The Exploration of Space* (Temple Press, 1951)

—— *A Fall of Moondust* (1961)

—— *The Fountains of Paradise* (1979)

—— 'A Meeting With Medusa' (1971) in *The Best of Arthur C. Clarke* (1973)

—— 'The Fires Within' (1949) in *Reach for Tomorrow* (1956)

—— *Rendezvous With Rama* (1973)

—— 'Silence Please' in *Tales from the White Hart* (1957)

—— *Tales from the White Hart* (1957)

—— 'What Goes Up' in *Tales from the White Hart* (1957)

—— *2001: A Space Odyssey* (1968)

Clement, Hal, 'Whirlygig World' (1953) in *Astounding SF* June (1953)

—— 'Dust Rag' (1960) see *Where Do We Go From Here?* ed Isaac Asimov (1971)

—— *Mission of Gravity* (1953)

Conklin, Groff (ed), *Great Science Fiction by Scientists* (1962)

Cooper, Edmund, *The Cloud Walker* (1973)

Cooper, Ralph S, 'The Neutrino Bomb' see *Great Science Fiction by Scientists* ed Groff Conklin

Crispin, Edmund, 'Science Fiction' *Times Literary Supplement* **23** October 865 (1953)

Dahl, Roald, *Charlie and the Chocolate Factory* (1964)

Darwin, Charles, *Origin of Species* (Penguin, 1982)

de Bergerac, Cyrano, *A Voyage to the Moon* (1657) see *The Man in the Moone* ed F K Pizor and T A Comp (Sidgwick & Jackson, 1971)

Defoe, Daniel, *The Consolidator* (1705)

Devereux, Paul, *Earthlights* (Turnstone, 1982)

Dewohl, Louis, 'Religion, Philosophy and Outer Space' *America* **92** 420–1 (1954)

Dick, Phillip K, *The Man in the High Castle* (1962)

——*Do Androids Dream of Electric Sheep?* (1968)

Dickson, Gordon R, The Dorsai Cycle, see *The Final Encyclopaedia* (1984)

——*Mission to Universe* (1965)

—— *Time Storm* (1977)

Doyle, Sir Arthur Conan, *The Lost World* (1912) and 'When the World Screamed' (1928) see *The Complete Professor Challenger Stories* (Wordsworth, 1989)

Einstein, Albert, 'On the Electrodynamics of Moving Bodies', see *The Principle of Relativity* Albert Einstein *et al* (Dover, 1952)

Forward, Robert L, *Dragon's Egg* (1980)

——*Starquake* (1985)

——'When the Science Writes the Fiction' see *Hard Science Fiction* ed George Slusser and Eric S Rabin (Southern Illinois University Press, 1986)

Fraknoi, Andrew, *Universe in the Classroom* (Freeman, 1985)

Godwin, Francis, *The Man in the Moone* (1638) see *The Man in the Moone* ed F K Pizor and T A Comp (Sidgwick & Jackson, 1971)

Goldsmith, Maurice, 'Soviet Science Fiction' *The Spectator* 21 August 226–7 (1959)

Goswami, Amit, *The Cosmic Dancers* (Harper & Row, 1983)

Graham, E D, *Man's Great Future: from the 50th Anniversary Edition of the Christian Science Monitor* (Longmans Green, 1959)

Gregory, Charles T, 'The Pod Society versus the Rugged Individualists' *Journal of Popular Film* **1** 3–14 (1972)

Harrison, Harry, *Make Room! Make Room!* (1966)

——*Plague from Space* (1965)

Haynes, Roslynn D, *H.G. Wells, Discoverer of the Future* (Macmillan, 1980)

Heinlein, Robert A, 'All You Zombies' (1959) in *The Unpleasant Profession of Jonathan Hoag* (1959)

——Future History Series, see *The Past Through Tomorrow* (1967)

——'Let There Be Light!' (1930) in *The Man Who Sold the Moon* (1950)

——'Life-line' (1939) in *The Man Who Sold the Moon* (1950)

——*Magic Inc.* (1940) see *Waldo & Magic Inc.* (1950)

——*Citizen of the Galaxy* (1957)

——*Have Spacesuit—Will Travel* (1958)

——*Rocketship Galileo* (1947)

——'Shooting Destination Moon' (1950) see *Focus on the Science Fiction Film* ed William Johnson (Prentice-Hall, 1972)

——*Space Cadet* (1948)

——*Stranger in a Strange Land* (1961)
——*Time for the Stars* (1956)
Herbert, Frank, *Dune* (1965)
Highet, Gilbert, *People, Places and Books* (Oxford University Press, 1953)
Hirsch, Walter, 'The Image of the Scientist in Science Fiction: a Content Analysis' *American Journal of Sociology* **43** 506–12 (1958)
Homer, *Odyssey* (Penguin, 1970)
Hoyle, Fred, *The Black Cloud* (1957)
——*October the First is too Late* (1966)
Hoyle, Fred and Wickramasinghe, N C, *Diseases from Space* (Dent, 1979)
Irvine, Max, *Neutron Stars* (Oxford University Press, 1978)
Jarvie, Ian, *The Philosophy of Film* (RKP, 1987)
Jones, D F, *Colossus* (1966)
——*Don't Pick the Flowers* (1971)
Judd, Cyril, *Outpost Mars* (1952)
Jung, Karl Gustav, *Flying Saucers* (RKP, 1959)
Kaufmann, William J, *Universe* (Freeman, 1985)
Kepler, Johannes, *Somnium* (1634) see *Kepler's Dream* John Lear (University of California Press, 1965) and *Kepler's Somnium* Edward Rosen (University of Wisconsin Press, 1967)
——*Harmonicae Mundi* see *Great Books of the Western World* vol 16 (Encyclopaedia Britannica Inc., 1952)
——*Astronomia Nova* (1609)
Laumer, Keith, *Worlds of the Imperium* (1962)
Len, Stanislaw, *Solaris* (1961)
Lewis, C S, *The Abolition of Man* (1943)
——*The Lion, The Witch and the Wardrobe* (1950)
——*Out of the Silent Planet* (1938)
——*Perelandra* (1943)
——*That Hideous Strength* (1945)
Linde, A, 'The Inflationary Universe' *Reports on Progress in Physics* **47** 925 (1984)
Lowenthal, Leo, *Literature, Popular Culture and Society* (Prentice-Hall, 1961)
Lucian, *Icaromennipos* see *Lucian II* in the Loeb Classical Library (Harvard, 1954)
——*A True History* see *Lucian I* in the Loeb Classical Library (Harvard, 1975)
Malzberg, Barry, *Galaxies* (1975)
Marcuse, Herbert, *An Essay on Liberation* (Beacon, 1969)
——*One Dimensional Man* (Abacus, 1972)
McConnell, Frank, 'Rough Beast Slouching: a Note on Horror Movies' *Kenyon Review* **32** 109–20 (1970)
——'Song of Innocence: the Creature from the Black Lagoon', see *Hal in the Classroom: Science Fiction Films* ed Ralph J Amelio (Pflaum, 1974)
McDonnell, Thomas P, 'The Cult of Science Fiction' *Catholic World* **178** 15–18 (1953)
Merchant, Carolyn, *The Death of Nature* (Harper & Row, 1980)
Merril, Judith (ed) *England Swings SF* (1968)
Miller Jr, Walter M, *A Canticle for Leibowitz* (1960)
Milton, John, *Paradise Lost* (Penguin, 1980)
Morris, Michael, Thorne, Kip S and Yurtsever, Ulvi, 'Wormholes, Time Machines and the Weak Energy Condition' *Physical Review Letters* **61** 1446–9 (1988)
Murphy, Carol, 'The Theology of Science Fiction *Approach* **23** 2–7 (1957)
Newton, Isaac, *Mathematical Principles of Natural Philosophy* (Greenwood, 1962)

Nicholls, Peter (ed), *The Encyclopedia of Science Fiction* (Granada, 1979)
—— *The Science in Science Fiction* (Mermaid, 1982)
Niven, Larry, 'All the Myriad Ways' (1971) in *All the Myriad Ways* (1973)
—— 'Flash Crowd' (1973) in *The Flight of the Horse* (1973)
—— *The Integral Trees* (1985)
—— 'Neutron Star' (1966) in *Neutron Star* (1968)
—— *The Magic Goes Away* (1978)
—— *Ringworld* (1970)
—— *Ringworld Engineers* (1980)
—— *The Smoke Ring* (1987)
—— *Tales of Known Space* (1975)
—— 'The Theory and Practice of Teleportation' (1969) in *All the Myriad Ways* (1973)
—— 'Unfinished Story No. 1' (1970) in *All the Myriad Ways* (1973)
Niven, Larry and Pournelle, Jerry, *Lucifer's Hammer* (1977)
Nourse, Alan E, 'Brightside Crossing' (1951) see *Beyond Tomorrow* ed Damon Knight (1965)
Orwell, George, *1984* (1949)
Ovid, *Metamorphoses* trans. May M Innes (Penguin, 1955)
Peary, Danny, *Cult Movies* (Delacorte, 1981)
Pilgrim, John, 'Science Fiction and Anarchism' *Anarchy* **3** 361–75 (1963)
Phelan, J M, 'Men and Morals in Space' *America* **113** 405–7 (1965)
Plank, Robert, 'Communication in Science Fiction' see *Our Language and Our World* ed
    S I Hayakawa (Harper & Row, 1959)
—— *The Emotional Significance of Imaginary Beings* (Charles C Thomas, 1968)
—— 'Names and Roles of Characters in Science Fiction' *Names* **9** 151 (1961)
Plato, *Republic* (Penguin, 1970)
Poe, Edgar Allen, 'Hans Pfaall—A Tale' (1835) in *The Science Fiction of Edgar Allen Poe*
    (Penguin, 1976)
—— *The Masque of the Red Death* (1842) *in Tales of Mystery and Imagination* (Dent, 1985)
—— *Tales of Mystery and Imagination* (Dent, 1985)
Pohl, Frederik, *Beyond the Blue Event Horizon* (1982)
—— *Heechee Rendezvous* (1984)
—— (ed) *The Expert Dreamers* (1962)
Pohl, Frederik, and Kornbluth, C M, *The Space Merchants* (1953)
Pournelle, Jerry (ed), *Black Holes* (1978)
Riesman, David, *The Lonely Crowd* (Yale University Press, 1967)
Roszac, Theodore, *Where the Wasteland Ends* (Faber, 1973)
Rottensteiner, Franz, *The Science Fiction Book* (Thames and Hudson, 1975)
Rubin, Steve, 'Retrospect' *Cinefantastique* **4** 5–22 (1976)
Sapiro, Leland, 'The Faustus Tradition in the Early Science Fiction Story' *Riverside Quarterly*
    1 3–18, 44–57, 118–25 (1964)
Shaftel, Oscar, 'The Social Context of Science Fiction' *Science and Society* **17** 97–118 (1953)
Shallis, Michael, *On Time* (Burnett, 1982)
Shapin, Steven, and Schaffer, Simon, *Leviathan and the Air-Pump* (Princeton University
    Press, 1985)
Shapiro, Stuart L, and Teukolsky, Saul, *Black Holes, White Dwarfs and Neutron Stars: The
    Physics of Compact Objects* (Wiley-Interscience, 1983)
Shaw, Bob, *Other Days, Other Eyes* (1972)
—— 'The Giaconda Caper' (1976) in *Cosmic Kaleidoscope* (1978)
—— *The Palace of Eternity* (1969)

——'Small World' (1978) in *A Better Mantrap* (1984)

——*Shadow of Heaven* (1969)

Shelley, Mary, *Frankenstein: or, The Modern Prometheus* (1818)

Shortland, Michael, 'Science as Metaphor' *Ideas and Production* **1** 21–8 (1982)

——'Wonder Stories in Alienland' *Science as Culture* **4** 7–43 (1988)

Shute, Nevil, *On the Beach* (1957)

Silverberg, Robert, The *World Inside* (1971)

Slusser, George, and Rabin, Eric S (eds), *Hard Science Fiction* (Southern Illinois University Press, 1986)

Smith, E E, The Lensman Series, see *Children of the Lens* (1948)

——*The Skylark of Space* (1928)

——*Triplanetary* (1934)

Sobchack, Vivian, *Screening Space: the American Science Fiction Film* (Ungar, 1987)

Soddy, Frederick, *The Interpretation of Radium* (1909)

Sontag, Susan, 'The Imagination of Disaster' see *Hal in the Classroom: Science Fiction Films* ed Ralph J Amelio (Pflaum, 1974)

Stapledon, Olaf, *Last and First Men* (1930)

——*Starmaker* (1937)

Stewart, George, *Earth Abides* (1949)

Swift, Jonathan, *Gulliver's Travels* (1726)

Tolkien, J R R, *The Lord of the Rings* (1968)

Trilling, Lionel, *The Opposing Self* (W W Norton, 1957)

Updike, John, *Roger's Version* (1987)

Vallee, Jacques, *Dimensions* (1988)

Velikovsky, Immanuel, *Worlds in Collision* (1950)

Verne, Jules, *20,000 Leagues Under the Sea* (1870)

——*Around the Moon* (1870)

——*Around the World in Eighty Days* (1873)

——*The Clipper of the Clouds* (1886)

——*Five Weeks in a Balloon* (1863)

——*From the Earth to the Moon* (1865)

——*Journey to the Centre of the Earth* (1864)

——*Master of the World* (1904)

van Vogt, A E, 'Black Destroyer' (1939) see *First Flight* ed Damon Knight (1963)

——*Slan* (1940)

——'The Storm' (1943) in *The Best of A.E. van Vogt* (1974)

Vonnegut, Kurt, *Player Piano* (1952)

Walpole, Horace, *The Castle of Otranto* (OUP, 1982)

Weart, Spencer R, *Nuclear Fear: a History of Images* (Harvard University Press, 1988)

Weinbaum, Stanley G, *A Martian Odyssey* (1949)

Wells, H G, *Experiment in Autobiography* (1934)

——*The First Men in the Moon* (1901)

——*The Invisible Man* (1897)

——*The Island of Dr. Moreau* (1896)

——*The Shape of Things to Come* (1933)

——*The Time Machine* (1895)

——*The War of the Worlds* (1898)

——*When the Sleeper Wakes* (1899)

——*The World Set Free* (1914)

Whyte, William H, *The Organization Man* (Touchstone, 1972)

Wilkins, John, *The Discovery of a World in the Moone* (1638) see *The Man in the Moone* ed F K Pizor and T A Comp (Sidgwick & Jackson, 1971)

Williamson, Jack, *Seetee Ship* (1951)

Wyndham, John, *The Chrysalids* (1955)

—— *The Day of the Triffids* (1951)

—— *The Kraken Wakes* (1953)

—— *The Outward Urge* (1959)

FILMOGRAPHY

The films mentioned in the text are listed here in alphabetical order by title, followed by director (d), screenplay writer (s), production company and date of first release.

Alien d Ridley Scott, s Dan O'Bannon, Fox/Brandywine–Ronald Shusett Productions, GB (1979)

Amazing Colossal Man, The d Bert I Gordon, s Bert I Gordon and Mark Hanna, Malibu, USA (1957)

Andromeda Strain, The d Robert Wise, s Nelson Gidding, Universal/Robert Wise Productions, USA (1970)

Atomic Kid, The d Leslie H Martinson, s Benedict Freeman and John Fenton Murphy, Mickey Rooney Productions, USA (1954)

Attack of the Crab Monsters d Roger Corman, s Charles Griffiths, Los Altos, USA (1956)

Back to the Future d Robert Zemeckis, s Robert Zemeckis and Bob Gale, Universal/Amblin, USA (1985)

Beast from 20,000 Fathoms, The d Eugène Lourié, s Lou Morheim and Fred Freiberger, Warner Brothers, USA (1953)

Beast with a Million Eyes, The d David Kramarsky, s Tom Filer, San Mateo Productions, USA (1956)

Bed-Sitting Room, The d Richard Lester, s John Antrobus, Oscar Lewenstein Productions, GB (1969)

Beginning of the End, The d Bert I Gordon, s Fred Freiberger and Lester Gorn, American Broadcasting/Paramount Theatres, USA (1957)

Ben Hur d William Wyler, s Karl Tunberg, MGM, USA (1959)

Birds, The d Alfred Hitchcock, s Evan Hunter, Universal/Alfred Hitchcock, USA (1963)

Black Hole, The d Gary Nelson, s Jeb Rosebrook and Gerry Day, Walt Disney, USA (1979)

Black Scorpion d Edward Ludwig, s David Duncan and Robert Blees, Amex Productions/Warner Brothers, USA (1959)

Blade Runner d Ridley Scott, s Hampton Fancher and David Peoples, Warner Brothers/Ladd Co, USA (1982)

Blob, The d Irvin S Yeaworth Jr, s Theodore Simonson and Kate Phillips, Tonylyn Productions/Paramount, USA (1958)

Brain That Wouldn't Die, The d Joseph Green, s Joseph Green, Sterling Productions/Carlton, USA (1959)

Bride of Frankenstein, The d James Whale, s John L Balderston and William Hurlbut, Universal, USA (1935)

Bride of the Monster, The d Edward D Wood Jr, s Edward D Wood Jr and Alex Gordon, Rolling M, USA (1956)

Cabinet of Dr. Caligari, The d Robert Wiene, s Carl Mayer, Decla-Bioscop, Germany (1919)

Chain Reaction, The d Ian Barry, s Ian Barry, Palm Beach Pictures, Australia (1980)

China Syndrome, The d James Bridges, s James Bridges, Mike Gray and T S Cook, International Picture Corp, USA (1979)

Close Encounters of the Third Kind d Steven Spielberg, s Steven Spielberg, Columbia/EMI, USA (1977)

Colossus of New York, The d Eugène Lourié, s Thelma Schnee, William Alland Productions, USA (1958)

Conquest of Space d Byron Haskin, s James O'Hanlon, Paramount, USA (1955)

Coogan's Bluff d Don Siegel, s Herman Miller, Dean Riesner and Howard Rodman, Universal, USA (1968)

Crack in the World d Andrew Marton, s Jon Manchip White and Julian Halevy, Paramount, USA (1965)

Creature from the Black Lagoon, The d Jack Arnold, s Harry Essex and Arthur Ross, Universal, USA (1954)

Creature with the Atom Brain, The d Edward Cahn, s Curt Siodmak, Clover, USA (1955)

Curse of Frankenstein, The d Terence Fisher, s Jimmy Sangster, Hammer, GB (1957)

Damned, The d Joseph Losey, s Evan Jones, Columbia/Hammer-Swallow, GB (1961)

Day of the Animals d William Girdler, s William and Eleanor Norton, Film Ventures International, USA (1977)

Day of the Triffids, The d Steve Sekely, s Philip Yordan, Yordan, GB (1963)

Day the Earth Caught Fire, The d Val Guest, s Wolf Mankowitz and Val Guest, British Lion/Pax, GB (1961)

Day the Earth Stood Still, The d Robert Wise, s Edmund H North, 20th Century Fox, USA (1951)

Day the World Ended, The d Roger Corman, s Lou Rusoff, Golden State, USA (1956)

Deadly Mantis, The d Nathan Juran, s Martin Berkeley, Universal International, USA (1957)

Destination Moon d Irving Pichel, s Rip Van Ronkel, Robert Heinlein and James O'Hanlon, Universal/George Pal, USA (1950)

Dr. Strangelove: Or How I Learned to Stop Worrying and Love the Bomb d Stanley Kubrick, s Stanley Kubrick, Terry Southern and Peter George, Columbia, GB (1963)

Dune d David Lynch, s David Lynch, Dino de Laurentis, USA (1984)

Earthquake d Mark Robson, s George Fox and Mario Puzo, Universal/Jennings Lang/Mark Robson, USA (1974)

Empire of the Ants d Bert I Gordon, s Jack Turley, AIP/Cinema 77, USA (1977)

ET—The Extraterrestial d Steven Spielberg, s Melissa Mathison, Universal, USA (1982)

Fedora d Billy Wilder, s I A L Diamond and Billy Wilder, Geria/SFP, W Germany/France (1978)

Fiend Without a Face d Arthur Crabtree, s W J Leder, Producers Associates, GB (1957)

Five d Arch Oboler, s Arch Oboler, Columbia, USA (1951)

Flight to Mars d Lesley Selander, s Arthur Strawn, Monogram, USA (1951)

Fly, The d Kurt Neumann, s James Clavell, 20th Century Fox, USA (1958)

Flying Disc Man from Mars, The d Fred C Brannon, s Ronald Davidson, REP, USA (1951)

Forbidden Planet d Fred M Wilcox, s Cyril Hume, MGM, USA (1956)

The 4-D Man d Irvin Shortess Yeaworth Jr, s Theodore Simonson and Cy Chermak, Fairview Productions, USA (1959)

Frankenstein d James Whale, s Garrett Fort, Francis Edward Faragoh and John L Balderston, Universal, USA (1931)

Frankenstein Meets the Wolf Man d Roy William Neill, s Curt Siodmak, Universal, USA (1943)

Freaks d Tod Browning, s Willis Goldbeck and Leon Gordon, MGM, USA (1932)

Frogs d George McCowan, s Robert Hutchison and Robert Blees, AIP, USA (1972)

From the Earth to the Moon d Byron Haskin, s Robert Blees and James Leicester, Waverley, USA (1958)

From Hell It Came d Dan Milner, s Richard Bernstein, Allied Artists, USA (1957)

Frozen Dead, The d Herbert J Leger, s Herbert J Leger, Goldstar/Seven Arts, GB (1966)

Giant Claw, The d Fred F Sears, s Samuel Newman and Paul Gangelin, Katzman/Clover, USA (1957)

Golem, The d Paul Wegener and Carl Boese, s Paul Wegener and Henrik Galeen, UFA, Germany (1920)

Green Slime, The d Kinji Fukasaku, s Charles Sinclair and William Finger, Toei/Southern Cross Films, Japan/USA (1968)

Head, The d Bob Rafaelson, s Bob Rafaelson and Jack Nicholson, Columbia, USA (1968)

Hell is for Heroes d Don Siegel, s Robert Pirosh and Richard Carr, Paramount, USA (1962)

H-Man, The d Inoshiro Honda, s Takeshi Kimura, Toho, Japan (1958)

Hurricane, The d John Ford and Stuart Heisler, s Dudley Nichols and Oliver H P Garrett, Samuel Goldwyn, USA (1937)

I Married a Monster from Outer Space d Gene Fowler Jr, s Louis Vittes, Paramount, USA (1958)

Incredible Shrinking Man, The d Jack Arnold, s Richard Matheson, Universal International, USA (1957)

Invaders from Mars d William Cameron Menzies, s Richard Blake, Edward L Alperson, USA (1953)

Invasion of the Body Snatchers d Don Siegel, s Daniel Mainwaring, Allied Artists, USA (1956)

Invasion U.S.A. d Alfred E Green, s Robert Smith, Columbia, USA (1952)

Invisible Ray, The d Lambert Hillyer, s John Colton, Universal, USA (1936)

It Came from Beneath the Sea d Robert Gordon, s George Worthy Yates and Hal Smith, Columbia/Sam Gatzman, USA (1955)

It Came from Outer Space d Jack Arnold, s Harry Essex, Universal, USA (1953)

Just Imagine d David Butler, s Roy Henderson, B G De Sylva and Lew Brown, Fox, USA (1930)

Kingdom of the Spiders d John Cardos, s Richard Robinson and Alan Caillon, Arachnid/Dimension, USA (1977)

King Kong d Merian C Cooper and Ernest Schoedsack, s James Creelman and Ruth Rose, RKO, USA (1933)

Krakatoa, East of Java d Bernard Kowalski, s Clifford Newton Gould and Bernard Gordon, ABC, USA (1967)

Land That Time Forgot, The d Kevin Connor, s James Cawthorn and Michael Moorcock, Amicus, GB (1974)

Lord of the Flies d Peter Brook, s Peter Brook, Allen-Hodges, GB (1963)

Lost Continent, The d Michael Carreras, s Michael Nash, Hammer, GB (1968)

Lost World, The d Irwin Allen, s Irwin Allen and Charles Bennett, 20th Century Fox/Saratoga, USA (1960)

Madame Curie d Mervyn LeRoy, s Paul Osborn and Paul H Rameau, MGM, USA (1944)

Magnetic Monster, The d Curt Siodmak, s Curt Siodmak and Ivan Tors, United Artists/Ivan Tors, USA (1953)

Man in the White Suit, The d Alexander Mackendrick, s Roger MacDougall and John Dighton, Ealing, GB (1951)

Man they could not Hang, The d Nick Grinde, s Karl Brown, Columbia, USA (1939)

Man Who Fell to Earth, The d Nicholas Roeg, s Paul Meyersberg, British Lion, GB (1976)

Man Who Lived Again, The d Robert Stevenson, s L du Garde Peach, Sidney Gilliat and John L Balderstone, Gainsborough, GB (1936)

Man Without a Body d M Lee Wilder and Charles Sanders, s William Grote, Guido Coen/Eros, GB (1957)

Meteor d Ronald Neame, s Stanley Mann and Edmund H North, Paladium, USA (1979)

Metropolis d Fritz Lang, s Thea von Harbou, UFA, Germany (1926)

Mr Drake's Duck d Val Guest, s Val Guest, Daniel M Angel/Douglas Fairbanks Jr Productions, GB (1951)

Modern Times d Charles Chaplin, s Charles Chaplin, Chaplin, USA (1936)

Mole People, The d Virgil Vogel, s Laszlo Gorog, Universal International, USA (1956)

Monkey Business d Howard Hawks, s Ben Hecht, Charles Lederer and I A L Diamond, 20th Century Fox, USA (1952)

Night the World Exploded, The d Fred F Sears, s Luci Ward and Jack Natteford, Clover, USA (1957)

No Blade of Grass d Cornel Wilde, s Sean Forestal and Jefferson Pascal, MGM, GB (1970)

On the Beach d Stanley Kramer, s John Paxton and James Lee Barrett, United Artists/Stanley Kramer, USA (1959)

Phantom From Space d W Lee Wilder, s Bill Raynor and Myles Wilder, Planet Filmways, USA (1953)

Planet of the Apes d Franklin J Schaffner, s Michael Wilson and Rod Serling, 20th Century Fox/Apjac, USA (1968)

Plan Nine from Outer Space d Edward D Wood Jr, s Edward D Wood Jr, J Edward Reynolds, USA (1956)

Prophecy d John Frankenheimer, s David Seltzer, Paramount, USA (1979)

Quatermass Xperiment, The d Val Guest, s Richard Landau and Val Guest, Exclusive/Hammer, GB (1955)

Quatermass II d Val Guest, s Nigel Kneale and Val Guest, Hammer, GB (1957)

Quiet Earth, The d Geoff Murphy, s Sam Pillsbury, Don Reynolds, Bruno Lawrence and Bill Baer, Cinepro/Pillsbury, New Zealand (1985)

Quintet d Robert Altman, s Frank Barhydt, Robert Altman and Patricia Resnick, 20th Century Fox, USA (1979)

Quo Vadis d Mervyn LeRoy, s John Lee Malim, S N Behrman and Sonya Levien, MGM, USA (1951)

Return of the Fly d Edward L Bernds, s Edward L Bernds, Associated Producers, USA (1959)

Robe, The d Henry Koster, s Phillip Dunne, 20th Century Fox, USA (1953)

Robocop d Paul Verhoeven, s Edward Neumeier and Michael Miner, Orion, USA (1987)

Robot Monster, The d Phil Tucker, s Wyott Ordung, Three Dimensional Picture Productions, USA (1953)

Rocketship XM d Kurt Neumann, s Kurt Neumann, Lippert, USA (1950)

San Francisco d W S Van Dyke, s Anita Loos, MGM, USA (1936)

Satan Bug, The d John Sturges, s James Clavell and Edward Anhalt, United Artists/Mirisch/Kappa, USA (1965)

Silent Running d Douglas Trumbull, s Deric Washburn, Mike Cimino and Steve Bocho, Universal/Michael Gruskoff/Douglas Trumbull, USA (1971)

Silkwood d Mike Nichols, s Nora Ephron and Alice Arlen, ABC, USA (1983)

Sleeper d Woody Allen, s Woody Allen and Marshall Brickman, United Artists/Jack Rollins-Charles Joffe, USA (1973)

Solaris d Andrei Tarkovsky, s Andrei Tarkovsky and Friedrich Gorenstein, Mosfilm, USSR (1971)

Soylent Green d Richard Fleischer, s Stanley R Greenberg, MGM, USA (1973)

S.O.S. Tidal Wave d John H Auer, s Maxwell Shane and Gordon Kahn, REP, USA (1939)

Squirm d Jeff Lieberman, s Jeff Lieberman, Allied International/Squirm Co, USA (1976)

Stalker d Andrei Tarkovsky, s Arkady Strugatsky and Boris Strugatsky, Mosfilm, USSR (1979)

Star Wars d George Lucas, s George Lucas, 20th Century Fox/Lucasfilm, USA (1977)

Tarantula d Jack Arnold, s Robert M Fresco and Martin Berkeley, Universal, USA (1955)

Target Earth d Sherman A Rose, s William Raynor, Abtcon Pictures, USA (1954)

Teenagers From Outer Space d Tom Graeff, s Tom Graeff, Topor Corp, USA (1959)

Ten Commandments, The d Cecil B deMille, s Aenaes Mackenzie, Jesse L Lasky Jr, Jack Gariss and Frederick M Frank, Paramount/Cecil B deMille, USA (1956)

Testament d Lynne Littman, s John Sacret Young, Entertainment Events/American Playhouse, USA (1983)

Them! d Gordon Douglas, s Tad Sherdeman, Warner Brothers, USA (1954)

Thing, The d Christian Nyby, s Charles Lederer, RKO/Winchester, USA (1951)

Things to Come d William Cameron Menzies, s H G Wells, London Films, GB (1936)

This Island Earth d Joseph Newman, s Franklin Cohen and Edward G O'Callaghan, Universal, USA (1954)

Time After Time d Nicholas Meyer, s Nicholas Meyer, Warner/Orion, USA (1979)

Time Bandits d Terry Gilliam, s Michael Palin and Terry Gilliam, Handmade Films, GB (1981)

Time Machine, The d George Pal, s David Duncan, MGM/Galaxy, USA (1960)

Twentieth Century d Howard Hawks, s Ben Hecht and Charles MacArthur, Columbia, USA (1954)

2001: A Space Odyssey d Stanley Kubrick, s Stanley Kubrick and Arthur C Clarke, MGM/Kubrick, GB (1968)

2010 d Peter Hyams, s Peter Hyams, MGM–United Artists , USA (1984)

Twonky, The d Arch Oboler, s Arch Oboler, Arch Oboler Productions, USA (1953)

Warning Sign d Hal Barwood, s Hal Barwood and Matthew Robbins, Fox/Barwood-Robbins, USA (1985)

War of the Worlds d Byron Haskin, s Barre Lyndon, Paramount/George Pal, USA (1952)

Westworld d Michael Crichton, s Michael Crichton, MGM, USA (1973)

When Worlds Collide d Rudoph Mate, s Sidney Boehm, Paramount/George Pal, USA (1951)

Willard d Daniel Mann, s Gilbert Ralston, Cinerama/Bing Crosby, USA (1971)

Zardoz d John Boorman, s John Boorman, 20th Century Fox, GB (1974)

Zero Population Growth d Michael Campus, s Max Ehrlich and Frank de Felita, Sagittarius, USA (1971)

Zombies of the Stratosphere d Fred C Brannon, s Ronald Davidson, REP, USA (1952)

INDEX